MACRAMÉ MAGIC

6 BOOKS IN 1

Abigail Beake

IMPORTANT!!
BEFORE YOU START READING...

Remember that you can also Download and Print
Instructions with Extra Pictures
for all the Projects in this Book

Please **scan the QR code** or follow the link on Page 166
to access the Free Easy-Print PDF Guides

TABLE OF CONTENTS

BOOK 2
Advanced Macramé

BOOK 3
Macramé Home Garden & Plant Hanger

BOOK 1
MACRAMÉ FOR BEGINNERS

**Remember to check out your Easy Download
and Print Instructions with Extra Pictures
for all the Projects in this Book**

Please scan the QR code or follow the link on Page 166

Introduction
JOINING THE WORLD OF MACRAME

Welcome!

You are about to start a journey of discovery into the world of this ancient art.

Indeed – because macramé, though it's now popular as a hobby, is much more than that: it's a noble craft whose origins are lost in the mists of time.

If you are getting hooked on macramé but don't know where to start, here's a handy comprehensive guide that will take you step-by-step through the process of learning this remarkable art.

If you have already learned the basics, this guide will provide you with the opportunity to master the secrets of this age-old technique.

Or, more likely, you have heard of macramé before and were instantly intrigued, but so far you have never tried to make a piece because it seemed too difficult.

This book will help you complete your first project.

If you've already taken your first steps into the world of macramé, you'll discover a handbook full of information, tips, and advanced techniques explained in detail – which are not found in any beginner's guide.

Finally, if you are already familiar with crafting but have never tried macramé, then be warned that you are about to learn a unique art that is unlike anything you have done so far.

You are about to open a treasure chest full of invaluable advice on macramé, that will allow you to create wonderful pieces.

You don't set off on an adventure to an exotic place without all you need to enjoy your trip: passport, guidebooks, maps... Similarly, first, you will be presented with all the tools and practical information without which it would be very difficult to successfully carry out any project you want to accomplish.

Then, through a method of project selection, this book will turn you from a novice to an artist capable of creating artifacts of rare beauty and elegance.

You will experience a full immersion in the world of macramé.

You will learn the origins and history of macramé, making you feel part of an age-old tradition.

You will discover all the benefits of this noble art and become proficient enough to accomplish things you didn't even imagine you could possibly do. You will amaze not only yourself but also the people around you.

By browsing the Internet, you can find plenty of tutorials and projects for both novices and non-beginners. None of them, however, will enable you to learn the ropes of this craft.

When learning from web tutorials, the risk is to just mimic simple movements, without a clue as to what you are doing. Therefore, you will never grow independent. Besides, it would be like trying to build a house on swampy ground. Without solid ground on which to lay the foundations, the house will never stand.

This handbook teaches you the ABCs of macramé and the necessary method to successfully create unique artworks. Finally, as you will find in the book, it will change you as a person.

Inside, you will find step-by-step explanations of several projects, ranked by difficulty. This is meant to enable you to start from the basics and get better and better, step after step.

The invaluable information you'll learn will enable you to avoid the common mistakes all beginners make, which might cause you to give up macramé before you even start tying a knot.

But that's not all; if you have children – or love children – there is also an unmissable section dedicated to them. How to keep them entertained while teaching them the long-lost art of manual skills. You will be amazed at all the benefits macramé can have for children and teens.

You will also find a chapter dedicated to Christmas, to practice and create elegant and original crafts for one of the most beautiful times of the year.

And – why not! – once you get good at it, you may want to consider putting your talents to good use and maybe even start a small business by following the advice you'll find in the second book.

To recap, what you now have before you is the only guide that takes you by the hand from novice to expert, while also discussing in detail the best applications of contemporary macramé:

1. Macramé for Beginners
2. Macramé for Expert Knotters
3. Macramé Home Garden & Plant Hanger
4. Fashion Macramé
5. Macramé for Christmas
6. Macramé for Children

This (and much more!) awaits you inside this book...

... enjoy the journey and happy crafting!

Chapter 1

THE ANCIENT ART OF MACRAMÉ

About macramé: an age-old craft becomes popular

Macramé is the art of creating objects by weaving and knotting strings or ropes together.

In a nutshell, we can describe macramé as the art of creating through knots.

Apparently, it was initially invented as a solution to a problem – later you will learn what I am talking about.

At first, the interweaving of the strings had a merely ornamental function.

Over the centuries and as the technique evolved, these decorations became more prominent than the tunic or gown itself, and a great many other precious objects began to be created with macramé.

The Charm of an Age-Old Tradition

Before reaching you through this book that will enable you to master its secrets, throughout the centuries this ancient technique has been handed down from one person to another.

Macramé is more than a simple pastime – it's a combination of the history, skills, and culture of the diverse peoples who have added to it over the years.

Macramé has a magical, elemental appeal just for the fact that you can create beautiful lacework without the need for any tools (needles or hooks) other than your own fingers. Just by weaving and knotting together simple strings.

You will be the bearer of a craft that has spanned the centuries and reached people across the world. What you're about to learn has something sacred in it.

When and Where Was Macramé Born?

As previously mentioned, macramé has ancient origins. It was born from weaving.

When weavers finished a piece – such as a dress, or a tunic – they had a problem: there was a risk that the threads would unravel.

Therefore, the solution they adopted was precisely to knot the ends of the fabric to prevent them from unraveling.

They invented a system whereby the fringe strings (i.e., the loose ends of the fabric) were divided into groups. Braided and knotted together, these produced hemlines which, depending on their complexity, created patterns ranging from basic knotted fringes, to elaborate lacework.

In time, a simple solution to a practical problem became a full-fledged art, whose outcome turned out to be more precious than the textiles it adorned.

It is believed that 'macrame' comes from the Arabic word migramah, or 'fringe'. In the 13th century, Arabic weavers or 'macrameers' began securing the loose ends of woven textiles such as towels and shawls with decorative knots.

From Assyrian friezes of the 9th century BC, we can also learn that macramé was born in the Middle East.

One of the rarest finds to date – a tunic dating back to the 1st century BC – was found in 1978 by archeologists excavating the fortified city of Qasr Ibrim on the Nile.

The Islamic world is responsible for spreading the craft of macramé throughout the Mediterranean area.

Through the Arabs, it also spread to Spain, where unsurprisingly it is also called *fleco morisco* (Moorish fringe), and hence to France and other European countries.

Until it finally reached the court of Mary II of England at the end of the 17th century. The Queen relished macramé so much, that apparently, she would teach it herself to her ladies-in-waiting.

Another intriguing narrative claims that macramé arrived in Italy and subsequently circulated throughout Europe thanks to Genoese sailors in the 15th century.

As we shall see later, it is indeed true that this craft was greatly developed by seamen. In their downtime, they took to the practice of making macramé artifacts to sell or use as trading items.

In Italy, in the beautiful coastal region of Liguria lined with ports big and small, there is still a thriving school that keeps the tradition of macramé knotwork alive.

From the Ligurian coast, the emigrants spread macramé virtually everywhere in the world – especially in South America.

Yet another account relates that the first artifacts reached Italy and Europe through the Crusaders returning to their home countries from the Holy Land.

Unquestionably, from the 15th century onward this art became widespread among the convents and monasteries all over Italy and across Europe. The nuns then taught it to the local women, who would make full trousseaux for young brides. Not to mention church furnishings, of course.

Thanks to its appeal, as it passed from person to person, and from hand to hand, macramé spread all over the world, while at the same time expanding to include new patterns and motifs.

A Sailors' Craft: McNamara's Knotting

Not everyone knows that sailors have played a crucial role in disseminating macramé to the four corners of the globe.

Being very proficient in the art of knot tying already, they mastered this technique as a pastime while on long and dull sea crossings.

At first, sailors utilized macramé to craft nautical items such as hammocks, lanyards, rope ladders, and rope handles.

But soon they tried their hand at creating small handicrafts such as object holders, footwear, belts, bracelets, hats, and so forth.

Such items they would then barter or sell in the ports they visited, generating a thriving trade.

So intertwined are the fortunes of macramé with those of seamen, that in the ports where macramé artifacts were sold, they became known as McNamara's lace or square knotting.

One of the core techniques of modern macramé – the square knot – should be credited to them.

The Golden Age of Macramé

In the modern age, macramé reached a peak of popularity under Queen Victoria of England.

As previously mentioned, it was first introduced into England in the late 17th century, during the reign of Queen Mary II. The favorite pastime of Queen Mary herself was teaching macramé to her ladies-in-waiting. This sparked a real craze for macramé.

By the late 1800s, most Victorian homes were bound to feature some kind of macramé item. And it was no longer just lace to embellish clothing, but curtains, tablecloths, and bedspreads.

Between the late 19th century and the early 1900s, there was a major revival of decorative arts in all European countries and the United States. Such a revival was probably fostered by Art Nouveau aesthetics.

Unfortunately, the trend was halted by the tragedy of War. Particularly after World War II, all interest in macramé and most similar ladies' crafts seemed to die out – perhaps because they were considered too frivolous at a time when the main priority was reconstruction.

The 1970s

After the War, macramé looked doomed to oblivion... Until the 1970s, when it enjoyed yet another comeback.

This is when it encountered the hippy movement. The ethnic appeal of macramé fitted well with the desire to break free from a standardized culture and a society based on the consumption of mass-produced, environmentally unfriendly products.

The outcome of this match was a flurry of colors and diverse applications.

Indeed, macramé was no longer used solely for household decoration and linens. It was now possible to make anything with macramé: bags, sweaters, and even footwear.

1970s Must-Have Decor: The Macramé Owl

A real cult item in the 1970s was the macrame owl.

It is unclear how exactly this craze originated, but it's a fact that in the 1970s, macramé owls became a hugely popular home décor piece in America and beyond.

This notoriety might be related to the mascot chosen by the U.S. Forest Service in 1971: Woodsy Owl.

The cute little owl wore a green cap with a red feather just like Robin Hood and, in keeping with the proverbial wisdom of owls, urged people to refrain from causing pollution with the famous catchphrase:

"Give a hoot, don't pollute!"

In many cultures, owls symbolize wisdom and are bearers of good luck.

Besides, there is another fact worth mentioning. After Richard Nixon's famous 1972 trip to China, which helped to relax the relationship between the two countries and gave rise to a very flourishing trade, the Chinese practice of Feng Shui gained popularity in the United States.

Feng Shui teaches how to channel energy forces to create harmony between humans and the environment they inhabit. And the owl, besides being a totem animal for many cultures, holds a very important role in Feng Shui which gives it a precise position within a living space.

Today: The Rebirth of Macramé

Once again neglected (but never forgotten) from the 1980s to the present, macramé is now experiencing yet another renaissance.

Aided by the visibility gained on mainstream social networks by the work of accomplished fiber artists, the magic of macramé is winning everyone back.

If you can think it, you can do it... with macramé!

That's a bold claim implying that this craft has endless applications – so much so, that it is back in the limelight not only among craft enthusiasts but also among interior designers. To the extent that it has spawned a movement called *boho chic* – which we will discuss in detail in book four – featuring macramé tapestries, curtains, plant hangers – you name it!

Haute couture also values macramé patterns enough to feature them on the world's most important catwalks.

Macramé for Everyone

Most people think they are not gifted in handiwork. I bet you too have uttered one of these phrases at least once: *"I'll never be able to do it," "it's too hard," "I'm not cut out for it."*

Well, one of the keys to the popularity of macramé is just that: anyone can do it.

It requires no special manual skills to start with. Indeed, all you need to begin with is a good amount of patience.

Anyone can achieve a good degree of proficiency in this art without any particular background skills, provided they have the patience to learn, the willingness to stick at it, an eye for detail, and attention to the steps to follow.

Plus, it is affordable and well within the reach of almost any budget.

Finally, it is environmentally friendly. Indeed, it uses natural fibers, and no scraps are wasted.

The Magic of Macramé: an Art No Machine Can Replicate

Nowadays, machines can replicate many varieties of lacework.

However, the focus of industrial manufacturing is not the beauty of the artifact, but mass production in ever-diminishing turnaround times.

No machine can replicate the accuracy and elegance of the human gesture creating a handcrafted one-of-a-kind piece.

This is where all the magic of macramé lies – a delicate and refined art that stems from a combination of creativity and know-how.

Only by this fusion can infinite patterns be created using just a few basic knots. The only limit to the ever-changing variety of shapes and objects is imagination.

What You Can Do with Macramé: 4 Macro Categories of Objects

As I just mentioned, when it comes to macramé the only limit is imagination. Indeed, there are countless items you can create by learning this wonderful technique. Here are some examples, divided into 4 broad categories:

1. **Home furnishings**

- curtains
- tapestries
- hanging planters
- mirrors
- household linens
- coasters
- vase covers
- lampshades
- rugs

2. **Jewelry & Accessories**

- pillows
- bags
- earrings
- bracelets
- belts

3. **Home & Garden**

- plant hangers
- lanterns
- baskets

4. **Special Items**

- dreamcatchers
- animals
- wreaths
- leaves.

Chapter 2
WELLBEING AND MACRAMÉ

Knots as Sacred Symbols

Knots have always been very powerful symbols for humanity. Not only do they represent connection, unbreakable union, and bonding, but they are also symbols of strength.

In ancient Egypt, the knot – as a magic symbol – represented the point where human forces converged with divine forces.

In ancient Indian tradition, the *infinite knot* is the symbol of a good omen. This is a closed pattern composed of lines intertwined at right angles.

Having neither beginning nor end, it perfectly represents our spiritual journey, not subject to time or space.

You must surely have seen in many handiworks and jewels the figure of the triple knot, or *Triquetra*. You may also know that this is the symbol of an ancient Northern European people: the Celts.

The word *Triquetra* means "triangle" and it is formed by three intertwining arches.

Being characterized by an unbroken line, it represents unity and eternal life.

The *Pan Chang* knot is one of the eight most important symbols of Buddhism. In Chinese, Pan Chang means "infinite" and represents the life cycle, without beginning nor end. The pronunciation of Pan Chang also means "happiness" which is why it is often given as a lucky charm.

As you can see, working with knots also has a strong spiritual aspect, which we shall further explore later in the book.

Have Trouble Meditating? Learn How to Macramé

When you start working with knots, you will find that the world around you will simply disappear.

No worries, no anxieties, no people around to stress you out with their demands. Just your hands moving in the present moment.

By learning a few basic knots, the design combinations you can create will be endless – allowing you to lose yourself in your creativity.

Besides letting you experience the thrill of wonderful patterns emerging out of nowhere as a result of your work, macramé can be a great form of meditation.

Working with macramé creates a magical vibe around you. You will greatly revalue the phrase, "happiness lies in the little things of life."

Macramé is a great way to gather your thoughts, root your energy, and be more aware of – and better in tune with – the present moment.

Our society has now reached levels of anxiety and stress that are increasingly difficult to manage. That's why meditation has become so popular.

But not everyone is suited for meditation. Not everyone can sit, cross their legs, and breathe, without starting to lose patience!

Yet the benefits of meditation are widely proven by countless scientific studies. So, what should we do, just forego the possibility of managing anxiety and stress without medication?

Sometimes even guided meditation won't help, because it's too easy to get distracted just from hearing a voice speaking in our ears. Besides, how can we choose the guided meditation that is right for us, from among the overwhelming choice of podcasts, apps, and YouTube videos?

Mindfulness – meditation experts say – is the only key to soothing the mind and managing external concerns.

However, a great many people feel the need to pursue mindfulness in more active ways.

The art of tying knots offers a real anchor to a restless mind overwhelmed by a thousand thoughts.

When you create macramé, your mind focuses on one single active pursuit. It doesn't have time to compulsively analyze all you have done in the past.

Through macramé, the mind is refreshed because it does not burn up unnecessary energy by constantly thinking about what you are going to do in the future.

How to Meditate and Relax with Macramé in 4 Steps

Forget awkward postures, mats, yoga bricks, and unnatural breathing that you will never be able to stick to on your own.

Meditating and relaxing are much easier than what they've always told you.

Here's how to take control of your mind, calm it down, focus on one task, and relax all else in 4 very simple steps:

1. Put on some music that you like and find relaxing, making sure the volume isn't too loud (in the second book, you'll find a section specifically devoted to choosing a suitable playlist)
2. Sit down where you usually work at macramé (in the next paragraph you will find all the instructions to create your macramé workshop) and grab your string
3. concentrate on the feeling of the string in your hands
4. practice the knots by always thinking only about the next move.

This method wards off distraction and prevents your mind from drifting and being overwhelmed by projects and worries (as a result of the first two steps).

And the great thing about working with macramé is that it's so minimal, so basic. You don't need any tools other than a piece of string and your hands. And you don't necessarily have to aim for a complicated goal.

You needn't be following a predetermined template. You just need to master the basic knots, and then allow yourself to let go. Without thinking. Let your imagination run wild and explore the wonder of watching wonderful art materialize out of thin air.

Nurture Your Body and Mind with the 7 Benefits of Macramé

The moment you start to macramé, you can immediately enjoy the powerful benefits of this age-old craft. There are so many of them, but here let me just list the top 7:

1. **It provides mental clarity**

As we saw in the previous paragraph, macramé not only relieves you from stress but also helps you to strengthen your concentration skills.

At work, do they want you stressed out and multitasking? Focusing on a single task to complete will bring you a previously unknown sense of calm, accompanied by positive feelings of well-being.

Your movements will create a peaceful atmosphere where you are the one deciding what to do – the one in charge of the game.

With your mind free of restlessness and focusing on specific movements, you will no longer be preoccupied with stressors, and will finally feel free to engage in super positive feelings for your mind and body.

2. **It boosts your brain activity**

Macramé is the perfect blend of technique and creativity. This means that when you practice it, you stimulate both brain hemispheres at the same time: the right one, more related to manual skills, art, and emotions; and the left one, which helps you learn skills and techniques.

As a result, you will be able to boost your ability to concentrate, think quickly, and learn.

In addition, as you acquire a new skill, you create new neuronal bonds and strengthen existing ones.

3. It increases your self-esteem and confidence

Holding something made with our own hands – setting yourself a task and seeing it accomplished – is a great way to boost your self-confidence and self-esteem.

Picture the sense of pride you will feel once you hold your first macramé piece in your hands, or a particularly challenging project that you managed to complete all by yourself.

4. It makes your fingers smarter

Learning how to make knots improves your motor skills because it develops the muscles of your fingers, hands, and arms. It helps loosen your joints and makes your fingers smarter.

It certainly improves your dexterity – and the more you practice, the faster your speed of execution will grow.

5. It makes you learn new skills

Did you know that many people who previously knew nothing about macramé have become so fond of this craft, that they are even setting up small businesses around it? (We shall see how in the next section of this handbook series)

In these challenging times, acquiring a new skill can be of great help should we ever face financial difficulties.

6. You challenge yourself while also having fun

Stop the "I'm not good enough," "I can't do it," and "it's too hard," attitude. Each new project you undertake will be a challenge to yourself and the negative part of yourself that's always saying you aren't equal to the task.

Silencing it will be such fun!

7. It improves your social relationships

As we have seen, macramé improves the quality of your life. Not only through its benefits for your mind and body. When they see you calmer, happier, and more relaxed, even those around you will benefit from the positive emotions you are experiencing. They will want to be around you, to spend time in your company. And you will be surrounded by the affection of family and friends.

Making room for your creativity will help you feel more accomplished... Think of how impressed they'll be when they get to see the artwork you made with your own hands.

A Healing Art

Calmness, relaxation, concentration and a sense of well-being have a very positive impact on both the body and mind and they significantly improve your mood.

Macramé works wonders for those suffering from anxiety, depression, or ADHD. Such people will find it a powerful tool to help them cope with their symptoms.

The calming effects of macramé trigger a flow of positive emotions and trigger the body to release beneficial chemicals such as dopamine.

As a result, a less stressed body is also able to perform all its vital functions to the fullest.

Chapter 3

WHAT DO YOU NEED? ALL YOU NEED TO KNOW BEFORE YOU START

Boards

The best way to do macramé is to anchor the strings to rigid supports so that they do not move as you tie the knots.

These boards should not be heavy, but light enough to fit them on your lap or carry them around.

As we shall see in detail in the next paragraphs, some are also equipped with grids and diagrams to guide pattern execution and measurements.

- **Rigid board**

all you need is a rigid but lightweight board to anchor the strings while you knot them. You can make it yourself with a cork or polyurethane board or you can buy them in stores (macramé boards)

- **Macramé board**

rigid support with grid and numbered notches to anchor the ropes (how to use the board is explained in detail in the next paragraph)

- **Pillow**

round or semi-spherical shape, useful for macramé lace. The sheet with the lace design is pinned onto the support

How to Use the Macramé Board

Macramé boards are lightweight boards featuring a measurement grid that you can use as a guide to ensure that the work is perfectly symmetrical.

Some boards also allow you to anchor the strings so that they serve as a frame. Other boards require drawing pins, ball pins, or T-pins to secure the string.

Most boards have a grid printed on the front. The grid is usually marked with inches or centimeters to give you an idea of the actual size.

Some boards also provide patterns of basic knots. This may be useful for novices who require constant referencing.

The grid is marked with numbers. The distance is measured in inches.

Macramé Tools and Accessories

You only need two things to get started: string and your hands.

But a few simple tools can also be very helpful.

Here you will find the main tools required for most projects. Other tools or accessories may vary depending on the design that you plan to make.

- **Macramé cord/rope**

Cords are the basic supply for macramé, the material out of which every macramé project is made. Read on to find a chapter devoted entirely to providing you with a complete guide to all the types of cord and rope that you can use.

- **Scissors**

You don't need fabric scissors; a pair of standard scissors will do. But they must be sharp enough to cut the cords easily.

- **Cutter**

Standard or rotating cutter. Required to cut thicker cords.

- **Measuring tape**

Tailoring or retractable. Useful to measure the length of ropes.

- **Fringe comb**

There are countless types. Their purpose is to separate and align the fringes.

- **Thread burner**

A thread burner is a marker-shaped tool with a metal tip that heats up. It is used to remove any excess string that interferes with the process. It works on both natural and synthetic fibers.

- **Pliers**

Here are the main types of pliers you might find useful:

- Flat-nose pliers, an essential tool if you plan to start making macramé jewelry
- Diagonal cutting pliers, for cutting wires and cables diagonally
- Cutting pliers, for cutting yarns and thin metal wire
- Round nose pliers, to work with metal wire without denting or ruining it; also used to finish the string ends or resize them
- Curved nose pliers, ideal to reach difficult spots and to open and close rings and chains
- Nylon jaw pliers, with coated jaws so as not to damage the ropes you use to make macramé.
- **Sewing Pins**

Useful to mark the length of braids and knots and to hold beads and other embellishments in place.

- **Map-style pushpins**

It's the type with the plastic head. Perfect for pinning your strings, especially on materials such as cork.

- **Long embroidery needles**

To secure the ends of your macramé and prevent it from coiling.

- **Tapestry needles**

Blunt needles with a big eye. The large eye is useful because it can accommodate threads or fibers thicker than the common sewing thread would. Most tapestry needles are large enough for crochet yarn or embroidery thread, and many are large enough to hold string as well. The largest needles can accommodate thicker yarns.

- **Macramé T-needles**

T-head needles are used with a macramé board to keep the threads in place while working.

- **Wool needles**

Blunt-tip needles with a large eye for medium and thick wool yarn.

- **S-hooks**

To hang macramé plant hangers.

- **Beading awls**

Useful tools to make the strings pass through bead holes. Simply thread the bead, with the end of the thread create a loop around two fingers, tuck the end of the thread inside the newly created loop, insert the tip of the spike inside the loop, then gently pull the end of the thread by bringing the knot closer to the bead.

- **Beads**

They come in plastic, ceramic, metal, in all sorts of shapes or colors – the only limit is the imagination.

- **Jump rings**

Small metal wire rings with a cut or split. They are used to attach charms, beads, clasps, or other items to a design or pattern.

- **Pendant connector rings**

Wooden or metal rings used as fasteners, to finish bags, for interior decorations, garlands, jewelry, necklaces, bracelets, ornaments, knitwear, etc.

- **Metal rings**

Bare rings of various sizes for dream catchers or garlands.

- **Macramé cord ends**

Steel or plastic bar or ribbon terminals, cones, and caps. There are many styles to choose from depending on the look you want to achieve for your project. Useful for key rings, belts, bracelets, bags, etc.

- **Wooden beads**

Wooden balls with holes. They come in different sizes. Ideal for making plant hangers, bags, jewelry, tapestries, and many other projects.

- **Wooden sticks**

Dowels of different sizes depending on the size of the tapestry you are making.

- **Glue**

You will mostly need glue to make macramé jewelry. They are particularly versatile adhesives sold in tubes equipped with a precision tip applicator ideal to apply small dabs without smearing the handiwork.

- **Epoxy resin**

Often used to create and decorate artistic objects. If you want to use it but don't know how – then I recommend experimenting at first with small amounts for small objects. There are also molds available on the market to create any type of object upon which to attach your macramé.

- **Stones and other materials**

To embellish your macramé pieces and to make jewelry, bags, and more, you can also decide to use stones.

Other materials may be used depending on the project you want to create.

Set Up Your Own Macramé Workshop According to the 7 Rules of Feng Shui

Feng Shui is a very ancient Chinese discipline that seeks to promote harmony and good circulation of energy between people and the environment they inhabit.

Accordingly, it addresses the layout of rooms and the basic furnishing elements within each of them.

Feng Shui means 'wind and water,' and it pursues the harmony that is achieved through the balance between the two opposing forces governing the universe, the Yin and the Yang.

Also important in achieving this balance are the 5 elements – Wood, Fire, Earth, Metal, and Water – out of which everything is made up, according to Chinese tradition.

The main principle of Feng Shui is that in every environment there are energies, so-called *Ch'i*, that positively interact with people, by stimulating them, soothing them, and inspiring optimism, courage, and self-confidence.

However, if something stands in the way of these sources, the energies are blocked.

Here are 7 tips to create the ideal environment for your macramé workshop.

1. If the walls of the room are white, make sure to have dark-colored objects in it
2. Natural light is preferable to artificial light, so choose a bright room
3. Place natural elements such as plants, stones, pebbles, or pieces of wood into the room.
4. Use a pink Himalayan salt lamp.

5. Scent the room by burning candles or incense sticks with natural aromas.
6. To let the energy flow naturally, it is useful to hang mirrors on the walls.
7. Use macramé objects you made as decor elements. Wall tapestries and plant hangers are particularly suitable for this purpose.

Working in a perfectly balanced environment will make you feel more serene. These "positive energies" will also be transferred to the objects that you create, giving them a special charm and magic that can't be found in commercial mass-produced objects.

Build the Perfect Macramé Workstation in 8 Steps

Here's how to design and build your perfect workstation. This section is particularly useful when you plan to make vertical projects such as tapestries or plant hangers. Let's get started!

1. Metal clothes rack

This is one of the most cost-effective choices for you to carry out any vertical project. You can use a standard clothes rack, preferably one with casters so you can move it wherever you want as needed.

My advice is to have two of them: one to support your work, and one to "store" the ropes you need.

2. Metal "S" hooks

Perfect to use with the coat rack to anchor your project in place while you're busy knotting.

3. Macramé brushes and combs

Another must-have tool for your workstation is a macramé brush or comb – preferably metal ones because plastic ones are too brittle, especially for thicker strings. They are essential to untangle small strands and fringes.

4. Macramé cord holder

When working on a project, you need to have all the necessary materials readily accessible. That's why you need to be able to store them neatly.

There are rope holders with hooks that fit nicely on clothes racks.

5. Metal Shelf

If you need a lot of string, it is useful to store the bobbins on a shelf.

6. Multi-purpose trolley

Ideal for storing items you need to have at hand such as pliers, brushes, needles, ropes, etc. The casters are very convenient to take it anywhere in the room.

7. Trash bin

As an alternative to the trolley, you may use a basket to store all the items you need as well as cord leftovers.

8. Desk or table

Some macramé projects are more easily carried out from a sitting position, with a flat surface for backing.

Where to Find the Best Supplies at the Lowest Prices

Earlier, we saw a comprehensive list of the supplies you may need to carry out your work. Most of them can be easily found online or in hardware stores, sewing suppliers, hobby stores, craft stores, and DIY centers.

Cords, ropes, and strings deserve a dedicated analysis. Below are all my tips to buy the best cords at the cheapest price.

Etsy

Currently, this is the most popular craft supply site. Regardless of your location, through Etsy, you can purchase supplies from sellers all over the world (who despite being small businesses, offer amazing quality). Here are some of the best Etsy stores that you can find by typing their name in the platform's search bar:

- Bobiny (Poland)
- MBCordas (Lithuania)
- UnfetteredCo (Canada)
- GANXXET (United States)
- Pepperell Braiding Company (United States)
- RockMountainCo (United States).

Amazon

This is the largest online marketplace in the world. Although it does not specialize in handmade products, you can find many sellers of macramé supplies

eBay

This is the best-known online auction site in the world. Over the years, it has become a great e-commerce platform. If you keep an eye out for good reviews, you can find excellent deals to buy cords, tools, and other supplies.

AliExpress

It is an online store platform owned by Alibaba, featuring small Chinese companies that offer products to international buyers.

Online specialty stores

In addition to the above-mentioned large platforms, there are also small specialty stores that sell online. These are often small businesses. Hence, prices may be a little bit higher, and shipping may be a little bit slower, but there will be no question as to the quality of the products you purchase.

Here are some of the best macramé e-commerce specialist stores:

- **HolmMade Macrame**

Founded by artist Angela Holm. The materials you find on this site are 100% natural and of the highest quality.

- **Modern Macrame**

Founded by Emily Katz, a true celebrity in the world of contemporary macramé. You will find a wide supply of cords and tools. Especially suitable if you love a more modern type of design.

- **Ganxxet**

Again, on this e-commerce site, you can find high-quality fibers and sustainable materials such as recycled cotton and organic fibers.

- **Niroma Studio**

Founded by Cindy Hwang Bokser, a fiber artist specializing in macramé. Suitable if you love unusual textiles and if you are particularly eco-conscious because many of her yarns are made from recycled materials.

- **The Lark's Head**

Founded by Rachel Breuklander, another successful fiber artist. In her store, you will find carefully selected fibers from all over the world.

Department stores and specialist retail stores

If you are not a fan of online shopping and you prefer to feel your supplies before you buy, the right solution for you is a brick-and-mortar store.

In the U.S. you can find macramé supplies in the following store chains (and although they are physical stores, they almost all have their e-commerce platforms too):

- Hobby Lobby
- Lowe's
- Home Depot

- Blick Art Materials
- Michaels
- Joann
- Walmart.

The only drawback could be the uncertainty regarding cord quality (in the next chapter, you will find an in-depth analysis on how to tell the quality of cords) and higher price tags.

For the highest quality, my advice is to look for specialty stores near you and read the reviews they have.

How to buy wholesale and save

If you are a passionate macramé enthusiast already and anticipate needing lots of cord, the cheapest solution might be to bulk buy.

You have two main options:

- AliExpress

As mentioned above, it ships worldwide directly from Chinese manufacturers.

- XKDOUS

This is an American company that allows buyers to purchase large quantities of rope with considerable savings.

The only drawback is that shipping can sometimes take a few weeks.

Chapter 4

TYPES OF CORDS: THE ULTIMATE GUIDE

Macramé: an Overview of the Materials

In principle, any suitably flexible and durable knotting material can be used for macramé.

Indeed, macramé cords can be made of many different materials. The first broad distinction is between:

* **Natural fibers** such as cotton, jute, wool, silk, hemp, linen
* **Synthetic fibers** such as nylon, acrylic, paracord, polypropylene, and plastic

As a rule of thumb, natural fibers are preferred for indoor projects and accessories, while synthetic fibers are more widely used for outdoor projects because of their greater durability.

The Top 3 Most Used Types of Cord

Regarding their structure, you should be aware that 3 types of ropes are best suited and most commonly used for macramé:

1. **twisted rope**
2. **string rope**
3. **braided rope.**

When carrying out projects it is not mandatory to use only one type of rope- on the contrary, you can have fun mixing them.

Twisted rope

Twisted rope is the one most used in traditional macramé.

It is made by twisting several threads (usually 2, 3, or 4) together into one single spiral.

This is highly recommended for beginners because it is extremely easy to work with.

The threads can be unraveled to make fringes that you can shape with a fringe comb.

Due to its particularly strong structure, it is also ideal for objects that have a real function, such as chairs or shopping bags.

On the contrary, it is not recommended for projects that require a certain elasticity, such as swings or hammocks. Being stiff, the structure would be prone to breaking.

Out of the three types, this is the kind of rope that weighs the most, so be careful if you use it for clothing, belts, or purses.

It comes in a huge variety of gauges and colors.

String rope

The combed rope, or warp, is produced by twisting various strands of cotton on themselves.

Its structure makes it the most pliable of the three. It is very easy to make into fringes.

Ideal for making macramé feathers or leaves. Excellent for making plant holders – especially heavily patterned and elaborate ones – tapestries, bags, key rings, cushions... this is a multipurpose rope

However, it is not particularly recommended for novices, because the fact that it frays can be annoying when you are a beginner and you want to concentrate on practicing how to make the knots.

It is very soft and lightweight, so it is also suitable for apparel.

This kind of rope also does not have an elastic structure, so it is not recommended for objects whose use may involve tension.

It's the cheapest type of rope of the three.

It is also available in a variety of colors and gauges.

Braided rope

It consists of a group of interwoven cotton yarns.

It is the most recommended for beginners, being the easiest to work of the three.

However, it is unsuitable for projects with fringes, because it cannot be frayed due to its strictly intertwined threads.

Other types of rope: waxed cord

Waxed cord is generally thinner because it is used mainly in the creation of jewelry.

This yarn is covered with wax to make it more water resistant and to prevent fraying.

The wax coating can make the color darker, but there is a huge variety of colors available on the market.

How to Choose the Cord Thickness?

Contemporary macramé has no standard thicknesses.

Let's say that the minimum string diameter is never smaller than 2mm/0.08 inch.

GOLDEN RULE: Large projects = large thickness, small projects = small thickness

The most widely used cord thicknesses range from 3 to 5 mm/0.12" to 0.2", but it all depends on the type of project you want to achieve.

A.   As a rule of thumb, a 4 mm rope is ok for large pieces, while for smaller items such as necklaces and bracelets, 1.5 mm is better suited.
B.   3 mm ropes can be a good middle ground to make a wide range of items.
C.   Don't forget that if you want to use beads, buttons, stones, or other decorations, it is important to choose your cord thickness accordingly.

How Do I Know How Much Cord I Need?

One of the biggest doubts for beginners is how to measure how much rope is needed to carry out a project.

GOLDEN RULE (applies to most cases): The length of your rope should be 4 times the size of the project you want to make. If the rope is folded in half 8 times.

HERE'S A PRACTICAL EXAMPLE:

If you know that the total length of your project will be 60 cm (23.62") all you need to do is multiply 60 by 4, and you will have the length of each string

60 cm x 4 = 240 cm (94.49")

if your project requires the cord to be folded in half (e.g., lark head knot), then:

60 cm x 8 = 480 cm (188.98")

Here are some tips to estimate the length of the strings even more accurately. Just follow these instructions:

A.   decide on the size of the project and the number of strings you want to use

B. once you know the project size and number of strings, you can estimate how much material you need in total
C. if the project has loose knots or perforated motifs, multiply the longest area by 4
D. if the project has tight knots, multiply by 5
E. if it implies complex knots, multiply by 6
F. if the project has lots of knotworks, then err on the safe side, because you will need much more rope
G. 2-ply cords take up more length per knot, so you won't need as much cord
H. don't forget to allow for extra length in case the project has a fringe
I. when in doubt, cut more rope.

Keep in mind that if you choose to follow a pre-made pattern, in most cases it will include a guide to cord lengths and thickness.

6 Ways to Choose the Best Cord for Your Project

When you are a beginner, difficulties may arise even before you start working with macramé. That's why in this chapter we will address one of the most important issues that you need to know: how to choose the right cord.

In later sections, we shall discuss in detail the types of yarns that should be used depending on the project.

For now, all you need to know is that there are 6 basic factors to consider:

1. Material

The first thing you need to decide is what material your cord should be made of. In the past, macramé was made only from natural fibers such as cotton, hemp, or jute.

By contrast, today macramé patterns are created from a wide variety of materials. Including synthetic fibers such as satin rayon and nylon.

Among natural fibers, the best material for beginners is cotton; while among artificial fibers, nylon could be the easiest to unravel.

Indeed, as a beginner, you should consider that you will make some mistakes, and as a consequence, you should choose a soft material that also allows you to easily unravel the knots if you make a mistake.

2. Durability

As a beginner, it is important to know how to assess what kind of cord is best suited for a pattern, based on its strength. Indeed, for some projects, the knots and structure must be solid.

Specifically, if you need a sturdy cord, you should consider using ribbon, jute, leather, and nylon.

Usually, the durability of the cord depends on whether it was made by weaving or twisting.

3. Composition

Is it better to choose a braided or a twisted cord to start with? Quite often, macramé novices tend to buy braided cord because it is the cheapest and more readily available.

However, when you get down to practical matters, you will find that although it is very suitable to provide solidity to the knots and structure, it is not easy to untie.

On the other hand, a twisted cord unravels much more quickly.

4. Stiffness

Especially as a beginner, it's wise to choose a flexible cord rather than a stiff one. This is because it's more difficult to bend and twist it to achieve the shapes required by the pattern.

So, you need to check the cable width and make sure it suits the job you want to do.

For example, if you want to make a necklace, you should use a thinner string of flexible material, such as a cotton cord of 2 mm/0.08" at most.

5. Structure

It is also very important to choose the structure of your material according to the project. If you want to make a bracelet, you cannot choose a material that is too coarse – simply because it would cause skin irritation.

In the case of necklaces and bracelets, for example, you should avoid coarse wool or jute cords and opt for silk, cotton, satin rayon, nylon, or leather instead.

6. Thickness

To choose the thickness of the cord, always start from the project you have in mind. Is it bracelets? Necklaces? Or home decor, such as tapestries?

In the product sheet of the cord, you intend to buy, always check the thickness and whether there is information about the projects for which that cord is best suited.

In general, for smaller items and jewelry, never exceed 2 mm/0.08".

Q&A: The Answers to All Your Doubts

I'm a beginner, what rope should I use?

As previously mentioned, you can use many different materials based on your needs, but the best option for novices is always cotton cords, ideally the 3-5 mm (0.12"-0.20") size.

- They are the easiest to work with
- If you make mistakes, you can easily untie the knots
- Plus, they let you create stunning fringes.

What's the difference between rope and cord?

Rope is usually the 3-PLY (three layers) cord, while cord is generally identified with single-twist cords.

Are there any rules for choosing my cord?

Rather than rules, I would say there are 3 guidelines:

1. Aesthetics: keep in mind that your projects will look very different depending on the cord you choose.
2. Project type: Besides aesthetic considerations, what type of object are you going to create?
3. A pattern = a type of cord. If you are following a pre-made design, it's best to use a single type of cord to make sure your knots take up the same space and do not cause problems when following the pattern.

Do you still have doubts? Here's a tip: order several kinds of cords, experiment with them, and choose the one you like best.

I'm following a pattern but I want to change the cable, how can I do it?

If you are following a pattern but want to change the thickness of the suggested cord because you don't like it or because you have a leftover cord length of a different thickness, don't worry!

On the Web, you can find some handy comparative charts showing you how your cord length should change based on its thickness.

Can I dye my cord myself?

Of course! That is if the cord is made of natural fiber. Synthetic fibers, on the other hand, do not take dyeing very well.

Chapter 5

10 BEST TIPS FOR BEGINNERS

In this chapter, I have collected the best tips for beginners from true macramé artists.

Even though you may be thinking that you only want to pursue it as a hobby, it's still important that you learn macramé methodically to ensure success. What's better than following expert advice?

They are the only ones that will allow you to successfully and quickly learn the macramé techniques without running the risk of giving up before you even see your first piece come to life in your hands.

1. Check the position of your back

Get right away into the habit of making sure that your back is upright and comfortably supported. This will be especially helpful when you start working on larger projects that will require you to spend considerable amounts of time knotting.

You must find a convenient position. Now and then take a break and do some stretching exercises.

2. Take your time

Don't be in a hurry and allow yourself time to learn. It's no big deal if at first what emerges from your fingers is not so great. You only become great by making mistakes and not being shy to experiment.

3. Learn the basic knots first

As you read on, you will find detailed explanations of the main macramé techniques and basic knots that will enable you to create a variety of projects.

Even by mastering just one knot – the square knot – you will be able to create a wide range of items.

Another knot you absolutely must learn from the very start is the lark's head knot.

4. Always make sure you have quality scissors at hand

The quality of the scissors you utilize is crucial because sometimes you may need to cut a large amount of cord all at once.

In addition to the scissors, make sure you have a good fringe comb in your workstation (see the previous chapter).

5. It's easier to work with a suspended cord

Especially for tapestries and plant hangers. You can get a sturdy coat rack where you can hang your projects with metal "S" hooks – or you may want to hang nails on the top of door frames.

6. Use binder clips!

Most knots call for 4 rope lengths, but not all of them are your working rope.

For example, as you practice the square knot you need to anchor the free strings onto something (even a chopping board or a clipboard may serve the purpose) so you needn't worry about holding the tension of the strings you're not working on. Once you become an expert, you will be able to do without!

7. Always cut more cord than you estimated

In chapter 4, you learned the most effective method to estimate how much cord you are going to need – but as you have seen, there is no foolproof rule because each project has its pattern, so the risk of miscalculating lurks just around the corner. To avoid running out of cord, err on the plus side when it comes to cutting it. You can always use any leftover cord for smaller projects like tassels, fringes, and any other decorative elements.

8. To learn your knots, start with a thicker cord

In the beginning, use a thicker string/yarn, let's say 3 mm/0.12" or more. This can help with the task of learning the knots.

9. Start small

A smart way to approach macramé is to start with smaller projects. These can be plant holders, bracelets, bookmarks, or key rings.

10. Use what nature has to offer

Take a hike in the woods and collect any fallen branches you like. Take them home and leave them to dry. Remove the bark and sand them with sandpaper. There you have a beautiful support for your tapestries!

Chapter 6

THE 7 MOST COMMON BEGINNER MISTAKES

If you are a beginner, read this chapter very carefully. Here we shall examine together some of the most common mistakes encountered by novices when starting to design macramé pieces.

1. The cord is too short

One of the easiest traps to fall into is using cords that are too short.

Here is a method to decide how much cord the various knots will require.

A. Take a cord and tie it to your support, then choose one of the knots you have learned to make (see the chapter dedicated to knots)
B. With a marker, mark the beginning and end of the knot you've tied
C. Measure the knot when it's done
D. Undo the knot
E. Your cord now shows the marks you drew with the marker
F. Measure them with your measuring tape, so you know for sure how much cord you used to make the knot of your choice
G. This method is valid regardless of the complexity of the knots, just mark the beginning and end.

2. A badly fastened double half hitch knot

The distinctive feature of this knot is the lead cord running diagonally over the rest.

When making this knot, you should be very careful to hold the lead cord at the desired angle. If you hold the lead cord firmly and wrap the knotting cords around it, it becomes easy to make this knot without any problems.

Sometimes, though, you can tighten it wrong. Indeed, by pulling the knotting cord too much, you could warp the lead cord.

If this happens, simply pull back the lead cord to restore it to the correct angle and double-check that you did not make a mistake when tying your knot, but simply tightened the knot incorrectly.

3. The open diamond

When we set out to create a diamond – one of the most common patterns in macramé – we need to divide the cords we are going to use into two groups, to create two strands of double half hitch knots running in opposite directions.

What you will be tempted to do is grab the first cord of the first group and use it as a lead cord to make the diamond. However, the result you'll get at the end is a diamond with an open head.

If you happen to make this mistake, undo everything and proceed as follows: cross the two lead cords and first, create a double half hitch knot between them. At this point – and only once you have tied this knot – can you proceed with these two cords as knotting leads.

At the end of the pattern, you will see that your diamond now has a perfectly symmetrical shape, and its head is perfectly closed.

4. "Skewed" square knot

If you must tie a square knot under a diagonal line, you should be very careful to not make the flat knot "skewed".

The flat knot must always be perfectly horizontal. If you tighten it against the diagonal line while making it, the result will be a slightly crooked flat knot.

This may cause the work to look messy and skew your end result.

5. Pattern errors

When tackling a very large job, you may happen to grab the wrong cords while creating a net with an alternating square knot.

If you have two groups of threads held together with a flat knot, you could mistakenly take only one thread from the right and 3 from the left. It is very easy to detect the mistake immediately so long as you have few cords.

However, you need to be much more careful with larger, more complex jobs or you may run the risk, just when you think you're on the right track, of having to undo everything.

Always pay close attention, especially when making the alternating flat knot, so you always have the correct result.

6. Leading cord or knotting cord?

Mixing up the leading cord with the knotting cord is a very common mistake, especially when working with thinner ropes where errors may be less obvious.

This mistake occurs most frequently when you take a break from a job and then resume it.

In this case, the only solution is to undo it and correct the error.

7. One missing round

Always remember to wrap the double half hitch with 2 knots. This rule may seem very simple, but if you are making a very long bar knot you may lose count and have one cord make only one round instead of two.

It so happens that the knot with only one round will be weaker, and therefore can come undone very easily.

However, unlike other mistakes that require you to undo everything, in this case, it is enough to widen the space between the knots a little bit just to make up for the missing round of cord.

The opposite may also happen making the knotting thread do one round too many. In this case, the problem is that besides using up more cord in length, the extra round also takes up more space in width.

How to fix this?

Locate the point where the extra round is, and gently undo the knotwork up to that point.

Chapter 7

THE ABC OF MACRAMÉ: 9 TECHNIQUES AND BASIC KNOTS

Practice the 9 Macramé Basic Knots

Let's now dive into the thick of the technique. After setting up your perfect workshop according to the Feng Shui guidelines and creating the perfect workstation, you are now ready for the next step: mastering the basic macramé knots.

This chapter of the book will provide step-by-step tutorials for the top 9 basic knots of macramé. if you learn them right away, they will enable you to carry out a huge variety of projects.

Before reading the descriptions of the basic knotting techniques presented here, I recommend that you jump to the end of the book where you will find a Glossary of all the macramé terms you need to be familiar with.

Please note it's important that you learn these knots in the order they are presented to you.

KNOT #1: the Lark's Head Knot

The first knot you should learn is the Lark's Head knot.

Sometimes it's also referred to as Cow Hitch Knot.

This knot is very important because it attaches your macramé cords to an object, such as a dowel, branch, or anchor rope.

STEP 1

fold the rope in half and place the ring on the dowel rod.

STEP 2

run the rope loop around the rod and pull the two ends of the cord through the loop to tighten.

STEP 3

pull the two ends of the cord to tighten the knot.

KNOT #2: the Reverse Lark's Head Knot

This reverse knot is a Lark's Head Knot executed in reverse.

It's like a Lark's Head Knot seen from behind.

STEP 1

fold the cord in half and place the loop under the dowel rod.

STEP 2

bring the loop forward and pull the two cords through the loop.

STEP 3

tighten the two cords to form the knot.

KNOT #3: the Half Knot and Square Knot (Left)

The square knot is one of the most widely used and can be facing both to the right and to the left. To make it, you need two cords.

The HALF KNOT is simply half a square knot.

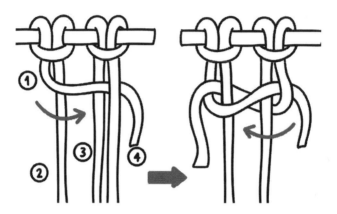

STEP 1

anchor two cords on a ring or dowel using the lark's head knot you just learned. As a result, you should have a total of 4 hanging cords.

STEP 2

create an inverted "D" shape with cord #1, the one in your left hand.

STEP 3

take the rightmost cord (4) and lift it over the tail of the inverted "D".

STEP 4

take the end of the right lanyard (4) and tuck it under the center of the support cords in the center (2-3), before pulling it back through the "D" shape (1).

STEP 5

pull both cords well.

So far, you have a half-square knot facing left.

STEP 6

now the position of your cord is reversed: working cord #1 is on the right while working cord #4 is on the left.

Take working cord #1 and move it left over the two filler cords and under working cord #4.

STEP 7

take the working cord #4 and gently move it to the right. Then go under the two filler cords and pass it over the working cord #1.

STEP 8

pull both working cords to tighten. Remember to keep the filler cords straight.

There you have the square knot facing left.

KNOT #4: the Half Knot and Square Knot (Right)

The process is the same, except it is reversed to the right.

STEP 1

create a "D" shape with the rightmost cord.

STEP 2

take the leftmost cord and lift it over the tail of the "D" shape.

STEP 3

take the end of the left-hand cord and pull it under the core filler cords, then outwards through the "D".

STEP 4

tighten the cords firmly.

By repeating this knot over and over again (in rows or sinnets) you will obtain a beautiful, knotted motif with a dense texture or mesh.

KNOT #5: the Alternating Square Knot

The alternating square knot is a knot used in many macramé projects, where it creates a great decorative pattern.

Alternating square knots are made by swapping the cords in use with each row.

To make an alternating square knot, start with 4 Lark's head knots (so you need 4 cords for 8 strands total).

STEP 1

for the first row, make the square knots you just learned how to tie.

STEP 2

in the second row, skip the first two strings before starting the new square knots.

STEP 3

in the third row, repeat the pattern of the first row.

STEP 4

in the fourth row, repeat the pattern of the second row, and so on.

STEP 5

make sure to tighten the ropes firmly without twisting them.

KNOT #6: the Spiral Stitch

The spiral stitch is also called "half knot spiral" because it consists of a series of half square knots.

This type of knot has the added quality of significantly embellishing the decorative pattern you are creating.

A spiral stitch requires at least 4 cords – 2 working cords and 2 filler cords – but more can be used.

Counting from left to right, mentally assign numbers 1 to 4 to these cords. Cords 1 and 4 are your working cords, while cords 2 and 3 are your filler cords.

The directions below are for the left-facing spiral knot. You can also start on the right side and make right-facing knots.

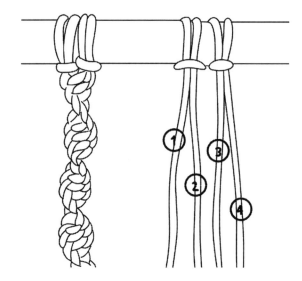

STEP 1

make a left-facing half-square knot. To do so, start with cord #1.

STEP 2

repeat the same knot repeatedly.

STEP 3

you will see that as you knot them, your cords will spontaneously curl into a spiral shape.

STEP 4

when knotting begins to get difficult because the cords twist upon themselves, just pull the cord and draw a full circle, then continue with your knots.

If you want your knots to face right, instead of starting with cord #1, start on the opposite side with cord #4.

KNOT #7: the Double Half Hitch (2 variants)

The "Double Half Hitch," also called "Clove Hitch," is one of the most widely used techniques in macramé because it adds beautiful designs to every project.

Here we present it in the two most frequent variants: diagonal and horizontal.

Sometimes (much more rarely, in fact) you might also encounter its vertical variant.

VARIANT #1: Diagonal Double Half Hitch

A.   LEFT TO RIGHT

STEP 1

take the cord on the far left (filler cord) and, holding it nice and tight, place it diagonally (towards the right side) over all the other working ropes.

Depending on how you position the filler cord above the others, you will have the direction and angle of your knots.

Step 2

take the following cord and try to bring it forward, then up and then around the filler cord you have on the left. Try to form a counter-clockwise loop.

Step 3

grab the next knot, pull it up, and tuck it through the loop. Pull to fasten a knot.

Step 4

repeat similar knots, always using the next working cord around the same filler cord.

Make sure your filler cord is always positioned diagonally. Each knot will be slightly lower than the previous one along the lead cord.

Step 5

once you have tied all the cords, you will get a full row of knots.

B. RIGHT TO LEFT

Step 1

take the cord on the far right (filler cord) and, holding it nice and tight, place it diagonally (towards the left side) over all the other working ropes.

Step 2

repeat the previous steps from the opposite side.

Step 3

once you have knotted all the cords, you will get a complete row of knots – in this case, each lower than the previous one starting from right to left.

VARIANT #2: Horizontal Double Half Hitch

Step 1

take the leftmost cord (lead cord) and place it horizontally above all the other cords (work cords) from left to right.

Step 2

make the half knots as before but in a horizontal line, going from left to right until you complete the row having knotted all the cords.

KNOT #8: the Overhand Knot

The basic overhand knot can tie multiple cords together. Very simple to make, this knot is crucial to start knotting.

Step 1

take the cords and bend them to form a loop

Step 2

tuck the end of the cords through the loop and tighten it.

KNOT #9: Gathering Knot

This knot is very important for making macramé plant hangers. Also called Wrapping Knot, it is a final knot that gathers all the other cords.

STEP 1

take a single cord (which will be your working rope) and form a long U-shaped loop over the bundle of filler cords you want to wrap. The loop should be facing down.

STEP 2

take the upper tail end of your single working cord (facing upwards) and wrap it around the filler cords and the U-shaped loop. Keep wrapping it down until the head of the slightly uncovered U-loop.

STEP 3

after wrapping the group of filler cords securely, thread the remaining working cord through the bottom of the loop.

STEP 4

carefully pull up the end of the tail that you threaded through the loop towards the top of the wrap. This will firmly secure all the wrapped cords.

STEP 5

give it a clean finish by shortening both ends of the working cord.

Chapter 8

HOW TO MAKE YOUR FIRST PROJECT IN 8 STEPS

You Are Now Ready for Your First Macramé Project

Now that you've set up the perfect workshop, sourced the materials to get started, and learned the basics of macramé knotting, you're all set to get started.

Below you will find step-by-step instructions to create one of the best projects for beginners.

I have chosen a mini version of one of the most beautiful objects you can make with macramé.

This will allow you to:

*   use less rope, to begin with
*   gain experience by starting small
*   practice the knots you've already learned
*   start with a delightful project that can be completed in a short time.

Before you begin, I suggest you should look at the glossary I have compiled for you. Thanks to this glossary, you will be able to use the key macramé vocabulary found in all the patterns.

DIY Mini Wall Hanger

What you need:

*   7 x 4ft Strings of 3mm Macrame Cord
*   scissors
*   fringe brush
*   wooden ring or dowel.

Knots Used:

*   Lark's Head
*   Square Knots
*   Double Half Hitch.

In this case, 5 cords were used to make 3 mini wall hangings

STEP 1

cut the cord you need. As we said, cut 7 cords of 4 feet each. You can follow our pattern or, if you feel confident enough, you can add cords as you like. The only limit is your creativity.

STEP 2

anchor all the cords to your ring or dowel of your choice with simple Lark's Head knots.

STEP 3

starting with the two central cords, create the first series of DHHK (diagonal half hitch knots).

This first series will run downwards from the center-left. A series runs downwards to the left. The other one runs downwards to the right.

STEP 4

starting again from the center, create another set of knots in a right downward direction.

STEP 5

start where you left off in STEP 3. Make another set of downwards DHHKs – but this time from left to center.

STEP 6

start where you left off in STEP 4. Make another set of downwards DHHKs – but this time from left to center. You created your first diamond pattern!

STEP 7

undo the cord ends left out of the knots with the brush to form a fringe.

STEP 8

you are ready to hang your first macramé project!

Square knot variant

STEP 1

cut 4 to 6 pieces of cord to a length of 4 feet.

STEP 2

create a series of alternating square knots that will form the first row.

STEP 3

keep creating as many rows as you like.

STEP 4

finish your piece by unraveling the cords with your brush to create the fringe.

Chapter 9

BONUS PROJECT

Your First DIY Plant Hanger

STEP 1

start with 4 cords. The length of the cords is up to you. Choose how big you want it to be and cut the cords a little more than twice the length you've decided upon.

STEP 2

fold the cords in half. You will end up with a total of 8 cords.

STEP 3

group all the cords together and form a single Lark's Head knot by threading the ends of the cords through the ring you have procured.

STEP 4

once the knot is secured to the ring, divide the cords into pairs.

STEP 5

decide at what height of the ropes you want your planter to hang. Then tie a simple overhand knot for each pair of cords you have created. Make sure the knots all fall at the same distance from the top ring.

STEP 6

place the joined cords on a flat surface and separate the cord pairs.

STEP 7

tie the loose end of one cord from each pair with a cord from the neighboring pair.

STEP 8

always tie the two cords with an overhand knot at a distance of 1-2 inches from the first knot.

STEP 9

divide the pairs once again and tie another series of overhand knots 1-2 inches further down.

STEP 10

gather all the strands 1-3 inches below the last knots and tie them together in a single overhand knot.

STEP 11

cut the leftover ends as long or as short as you like them.

STEP 12

insert your planter. Congratulations: you completed your first plant hanger!

LEARN THE MACRAMÉ LINGO

A Comprehensive Glossary for Beginners

A

Adjacent

Next to each other (knots or ropes)

Alternating

Usually, it is a term that is used with knots and it means that you tie a knot with a cord and then move on to the next with an alternating pattern.

ASK

Acronym for Alternating Square Knot.

B

Band

A large, flat length of macramé

Bar

A series of knots forming a raised area in the design. Half knots are often used to create bars and can run horizontally, vertically, or diagonally on a macramé piece.

Bight

A narrow section of folded rope which is threaded through other parts of the knot.

Body

This is the main segment of the project you are working on.

Braid

3 or more strings intertwined to form a braid.

Braided Cord

A type of cord made of 3 or more thinner strings braided together.

Bundle

A Group of cords gathered.

Button Knot

Firm round decorative knot.

BH

Short for Buttonhole. Lark's Head knots are used to make fastening loops.

C

Cord

Ropes of any natural or synthetic fiber used for macramé.

Core

Group of cords that run through the center of a project. Sometimes they are called filler cords/ropes or center ropes.

Crook

A cord section bent into a loop.

D

Diagonal

A cord or row of knots that runs from the top left to the bottom right (or vice versa). This is widely used in macramé patterns.

Diameter

The thickness of a cord. It is measured in millimeters or inches.

DDH

Short for Double Half Hitch. Made with two half knots tied side by side.

F

Fillers

Cords that run through the center of certain knots to support the design of your macramé. Also known as center cords.

Findings

Any fastening attachments, other than cords. They may be clips, metal or plastic bars, and attachments for accessories.

Finishing Knot

A type of knot that secures the ends of the cords to prevent them from coming undone.

Fringe

The ends of the cord are unraveled and left hanging to form the so-called "fringe", one of the most distinctive elements of macramé.

G

Gathering Knot

A knot that gathers the cords together and attaches them to each other, either at the beginning or at the end of a project.

H

Hitch

A type of knot that is often used to attach cords to other objects.

Horizontal Design

A design running horizontally or from side to side.

HSK

Acronym for Half Square Knot.

HHK

Acronym for Half Hitch Knot. HHKs almost invariably come in pairs, so you will see the DHHK abbreviation used most often in patterns. HHK also has diagonal, horizontal, and vertical variants. Thus, you can meet abbreviations like VHHK and HHHK.

I

Interlace

Patterns whose strings are intertwined to connect different areas of the design.

Inverted

Upside down from the starting position.

K

Knotting Cord

The cord used to create the knots of a design.

L

LHK

Acronym for Lark's Head Knot.

Loop

A loop is created when two parts of a rope intersect to form a circle or an oval.

M

Macramé Cord

Macramé cord generally consists of 6 intertwined threads. It can be in either natural or synthetic fiber and its main feature is resistance.

Macramé Rope

Typically made of 3 cotton threads twisted around each other. It is stronger than cotton cord and not as soft, but it's a great choice to make a plant hanger.

Macrame String

Super soft, single-twist rope, most commonly made of cotton.

Mount

The object on which the macramé is mounted. For example, it may be the wooden ring on which a plant hanger can be mounted.

N

Natural

This term is used regarding the materials of which the cords are made, i.e., natural fibers vs. synthetic ones.

Netting

A knot pattern that leaves many open spaces between the knots. It is often used to create plant hangers.

O

OH

Acronym for Overhand Knot.

R

RLHK

Acronym for Reverse Lark's Head Knot.

RHSK

Acronym for Right Half Square Knot. The opposite of HSK.

S

Scallops

Knotted loops of nodes created along the edges of the design.

Segment

A specific portion of area, cord, or pattern.

SK

Square Knot. A left half square knot and a right half square knot together form a full square knot. It is one of the basic knots of macramé.

Sinnet

Or Sennit. A row (or column) of identical knots.

Standing End

The cord end secured to the macramé board or any other surface to keep the cords tensioned.

Stitch

Sometimes used in place of 'knot'.

Synthetic

Synthetic fibers of which ropes are made, as opposed to natural fibers.

CONCLUSION

Now you're ready for the next step

Well done!

You managed to complete your first work all by yourself by following the ideas and instructions in this guide. I do hope you enjoyed your introduction to the engaging ancient art of macramé.

Now that your immersed in this creative activity, we will support you, whereas most other books will desert you or ask you to spend money to sign up for a webinar. We invite you to continue your journey with us and learn advanced macramé... Who knows, you too might become one of the successful artists you see on Instagram today!

BOOK 2
ADVANCED MACRAMÉ

**Remember to check out your Easy Download
and Print Instructions with Extra Pictures
for all the Projects in this Book**

Please scan the QR code or follow the link on Page 166

Introduction
YOU ARE NOW READY TO LEVEL UP!

As you read in the first Chapter of Book 1, covering the history of macramé, this is a textile art in which you can attain very high technical proficiency. At the same time, it allows you to explore your creativity to your great satisfaction.

The versatile nature of macramé has allowed it to adapt to the taste and fashion of many eras, including our own, as we shall see in the chapter on Boho Style macramé.

Besides, as you have already seen, endless items can be made from tapestries to plant hangers, from coasters to curtains, not to mention earrings, bracelets, necklaces, and virtually anything else you can think of.

If the first book was a full immersion in the world of macramé, in the second one the challenge becomes even more exciting.

After building your ideal workshop, setting up the perfect workstation, and getting everything you needed, you learned the main techniques and basic knots, which have already enabled you to undertake several projects.

How about stepping it up a notch? Are you ready to master the main advanced techniques and learn many professional tips to successfully identify and fix any kind of tricky issue with your macramé?

In this book, you will also find invaluable information, such as instructions for making a specific item that is very fashionable right now.

Or learn how to easily read patterns and make your own.

But that's not all!

If you browsed the table of contents, you may have seen that there is an entire chapter devoted to turning your passion into a business... Aren't you curious to learn how?

Chapter 1

TOOLS AND MATERIALS FOR ADVANCED PROJECTS

Advanced Tools

Crochet Hook

A useful tool to have when making macramé is a crochet hook. This is especially true when working with thicker cords.

That's because in such a case it may be difficult to thread the loose cord ends through the knots – an issue that can be solved with the help of a crochet hook.

It is also very useful to join two separate pieces of macramé together. For example, the two parts of a clutch purse.

Cord Dispenser

Another very useful tool when you need to cut the cords for your projects is a cord dispenser.

A cord dispenser consists of a base and two rods. One of the rods has a hole.

All you must do is place the cord spool on the dispenser, take the end of the cord and thread it through the hole on the other rod of the dispenser. Thanks to the dispenser, pulling the cord will no longer be an issue, and the cord will stay nicely tensioned and tidy without the risk of it becoming entangled.

When you simply cut it from the spool, your cord often tends to get entangled and coiled up, hindering you from smoothly continuing your work.

Cork Board & Pins

If you do many small projects such as jewelry – and therefore tend to use thinner thread – it is good to have a smaller work surface to work with.

My advice is that you should use a small cork board and pins (there are many pin cushions available on the market).

Pins are very handy because they allow you to anchor sections of the pattern so that you can work on it on the board.

Macramé Board Holder

In addition to the macramé board, you can also order a metal stand or easel with which you can tilt the board.

This allows you to have the cords at eye level and work more comfortably.

Tying Station

This tool – a sort of elongated clamp – helps you secure the ends of your cord when making a bracelet or a necklace.

This is very convenient if the patterns you are planning to do are particularly intricate.

Bead Threader

This is sort of a wand with a metal hoop at one end that helps you thread the beads on the cord quickly and easily.

A must-have nifty tool to avoid unnecessary stress, especially when working with small beads.

Wooden models should be preferred because they are sturdier than plastic.

Laser Level

Have you ever tied a series of knots that you couldn't align on the same level? Then you need a laser level!

A laser level is a useful device to correctly measure the slope of a surface in relation to a horizontal plane through a beam of laser light that projects a series of points belonging to a plane.

Macramé Separator

A strip made from wood or other material, of different sizes depending on the distance you want to achieve between the knots. With a separator, you make sure that the distance between two knots is correct. This gives your project a clean and tidy look.

Macramé Triangle

This will help you create beautiful home furnishings such as tapestries, dreamcatchers, plant hangers, and more. You can also combine them to create a striking 3D effect.

Macramé Hoop with Crossbar

Can be used to create tapestries, dreamcatchers, and plant hangers. As a pair, it can also serve as handles for a bag.

Macramé Bamboo Hoop

It can be used for hanging dreamcatchers, as a towel holder or to create the handles of a bag. Bamboo is a versatile, organic, and sustainable material.

Cord thickness chart made easy

A blanket rule that works well in lots of projects is to divide the cords according to 3 thicknesses.

Through this classification, you will be able to quickly determine the right cords for each project you want to make, out of endless options.

Always keep handy this simple table of thicknesses (and the caption below), and you will no longer be at risk of making mistakes:

Features	Small	Medium	Large
Size	1-3 mm/0.04-0.12 inch	3-7 mm/0.12-0.28 inch	7+ mm / 0.28+ inch
Project Type	earrings bracelets necklaces other types of jewelry other small projects	wall hanger plant hanger home decor other projects	large wall hanger other projects
Standard dimensions for specific projects			
As mentioned, in the beginning referring to standard measurements depending on the project you have in mind can be of great help:			
Big wall hanging	7+ mm (0.28+ inch) or 4-10 mm (0.16-0.39 inch) for more complex patterns		
Small wall hanging	3-5 mm (0.12-0.20 inch)		
Plant hanger	3-5 mm (0.12-0.2 inch)		
Macramé bag	3-5 mm (0.12-0.20 inch)		
Jewelry	1.5-3 mm (0.06-0.12 inch)		
Key chain / Bookmark	1.5-3 mm (0.06-0.12 inch)		

The 3 most used cord thicknesses

1. Small cords (1 to 3 mm/0.04-0.12 inch): jewelry and smaller projects

Cords of this thickness are often referred to as micro macramé cords. They are the thinnest macramé cords, and they are used to make virtually all types of jewelry and similar small projects.

This is because their thinner size allows you to execute the more complicated patterns. Additionally, these cords are thin enough to thread even smaller beads.

However, they are unsuitable for larger projects (such as plant hangers), not least because it would take forever to finish the project.

You would only want to use them to make a particularly elaborate pattern but do keep in mind that it will take much longer.

2. Medium cords (from 3 to 7mm /0.12-0.28 inch): medium-sized wall hangers and plant hangers

Medium cords are the most used for virtually any type of design.

Most likely, this is because their size makes them sturdy enough for wall hangers and plant hangers, as well as for other home decor items.

3. Large cords (7mm/0.28 inch up): medium and large wall hangers and plant hangers

Macramé cords of 7 mm (0.28 in.) or more are sure to have a great visual impact.

Due to their considerable thicknesses, making many knots with these cords will be difficult. Still, they are great if you want to make a tapestry big enough to cover an entire wall or a long railing, or even to create decorative curtains.

Some cords are so thick that you will need to use your arms as knitting needles, but the result may be striking.

Besides, the sheer impact of their size means you won't need to create particularly intricate and complex patterns.

Materials for advanced projects

Ramie

A textile fiber that is extracted from the bark of the ramie tree. It is a coarse, very durable, and beautiful fiber, and it provides a rustic finish.

If you are considering using it, protect your fingers with thin gloves or wraps to prevent blistering.

Sisal

Sisal is a fiber extracted from the leaves of the sisal plant which belongs to the agave family. Again, this fiber also adds a rustic touch to your projects, and as a bonus, it is easily obtained commercially in different thicknesses.

Again, protect your fingers before starting to work with it.

Polyester

If your project is meant for outdoor use or display, your best choice is not natural fibers but synthetic ones – such as polyester, which is much more durable.

Links & Connectors

When making your macramé jewelry, you may need findings and connectors. You can find all sorts of them in stores. They can also be used as design elements to embellish your bracelets and necklaces.

Very useful are the ones with 2 (or more) connection points to insert the finding into your pattern.

Oval Wooden Beads & Spacer Beads

Wooden beads with an oval shape are perfect for creating macramé bracelets, necklaces, or home decor items.

Swivel Lobster Clasps

Swivel lobster clasps are especially suitable to make purses, backpacks, handmade dog leashes, luggage straps, and so on.

Stones and macramé: meaning and symbolism of the top 10

Semiprecious stones are often used to embellish macramé jewelry.

They have always been considered to represent energy and purity. Gifting them is a gesture of love and friendship.

But you should be aware that choosing a stone is not a random decision, because they each have their individual meanings.

Let's explore the symbolism of the top 10 stones.

1. AMETHYST

Amethyst is the stone of introspection. Its name derives from the Greek Amethystos, which means "he who does not get drunk".

It represents the purity of soul, power and happiness, sincerity, humility, and wisdom.

Properties: it gives intellectual and psychic balance. It is considered the stone of introspection and protects against negative thoughts and "bad" intentions.

It eases the fear and helps one cope with situations more deliberately. Its purple color fights depression and stimulates spiritual development.

2. CORAL

Coral can be white, pink, or red and it only grows in clear, unpolluted, nontidal waters.

It represents sensual love and passion. It is considered a lucky stone.

Properties: it has a purifying and strengthening action. It stimulates energy and vitality. It eliminates energy imbalances due to negative moods. It opens the heart – so much so that it is considered the stone of affection.

3. PEARL

Pearls (and the mother of Pearl) symbolize perfection, uniqueness, and inner wisdom.

Pearls represent health. They are one of the main symbols of femininity. They evoke sensuality, authenticity, and beauty.

Properties: A pearl can produce calm and serenity by easing emotional tensions and anxiety. It fosters awareness and self-confidence. It protects love and friendship and attracts positive relationships and serene thoughts.

4. AGATE

Agate is the queen of protective gemstones – so much so that it is traditionally considered a powerful lucky charm.

It represents the balance between the physical and psychic world, will, and energy.

Properties: it boosts willpower. It counteracts negative influences. It promotes the mind-body balance. It favors soothing and relaxation. It provides sensitivity and tenderness. All types of Agate are in harmony with the vibrations of the earth and with feminine energies.

5. ONYX

The term *Onyx* is derived from the Greek word for "nail" or "claw". This recalls the hardness of this stone.

It represents inner strength and enthusiasm. It is also considered the stone of maturity.

Properties: it absorbs negative energies and protects against bad thoughts. It balances body-mind energies. It stimulates the intellect and increases concentration.

It relieves stress. It fosters harmony with others and increases self-esteem and self-understanding.

6. CARNELIAN

For ancient Egyptians, the carnelian meant "sun at sunset". Its orange hues embody feminine energies.

It represents joy, courage, and imagination.

Properties: it promotes vitality, optimism, and joy. It improves mood. It increases creativity and provides stability.

7. MALACHITE

Malachite comes from the Greek *malache*, meaning "mallow". It is also called the "mirror of the soul" and represents the ability to observe.

Properties: it increases the love for beauty and the thirst for knowledge. It improves expression and understanding, reduces fear, and removes selfishness by helping one to get in touch with others on an empathic level. It promotes critical spirit and vivacity of thought.

8. LAPIS

Lapis lazuli is one of the most famous gemstones since ancient times. Its name derives from the Persian *lapislazulus*, which means "blue stone".

It represents honesty and wisdom.

Properties: it stimulates intuition and concentration. It has a calming and regenerating effect. It helps with inner clarity and intentions. It supports the willingness to make decisions and promotes kindness.

9. ROCK CRYSTAL

Rock Crystal (or Hyaline Quartz) is considered the "stone of light".

It represents light, energy, and thought

<u>Properties</u>: it strengthens the ability to understand others and, at the same time, to affirm one's deepest nature. It enlightens the thought.

It stimulates self-awareness. It brings harmony, joy, and serenity. It helps with meditation. And finally, it is a powerful conductor of energy.

10. TOURMALINE

Tourmaline is the crystal of protection and luck.

It represents purification, endurance, and good luck.

<u>Properties</u>: it protects from negative energies and sad thoughts. It strengthens self-esteem and attracts luck.

Chapter 2

MAKE YOUR PATTERNS STAND OUT WITH 6 ADVANCED VARIATIONS OF THE BASIC KNOTS

After learning in the first book the basic macramé knots, such as the square knot, the lark's head, and the double half knot, you can already make a multitude of projects even without relying on "ready-made" patterns – simply by combining the techniques you have learned so far. This will provide an original touch to all your projects.

3 Square Knot Variations

1.   The Wave Knot

To do this, you need to have 3 cords (6 strands) attached to the dowel with lark head knots.

STEP 1: *Take the 3 cords on the left, then the one on the far right. The remaining two will be your filler cords.*

STEP 2: *Take the first 3 cords on the left, hold them together and bring them behind the two filler cords.*

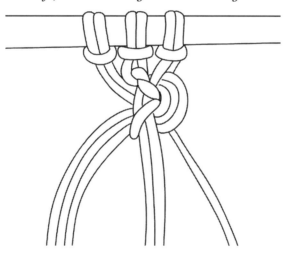

STEP 3: *Hold everything together with your left hand, while with your right hand you take the remaining free cord on the far left. Bring it behind the three left cords and in front of the two filler cords. Thread the cord into the small loop that formed in the middle.*

STEP 4: *Make sure you hold the three left cords together neatly, then tighten everything together.*

STEP 5: *Take the three cords (which are now on the right) and bring them back behind the two filler cords to the left.*

STEP 6: *You will still be holding your single cord in one hand; now, bring it back behind the 3 cords, in front of the filler cords, then thread it again through the small loop that has formed. Tighten everything with care.*

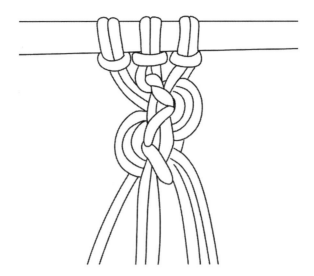

Step 7: *Repeat the entire sequence, only this time the 3 cords will be on the right again.*

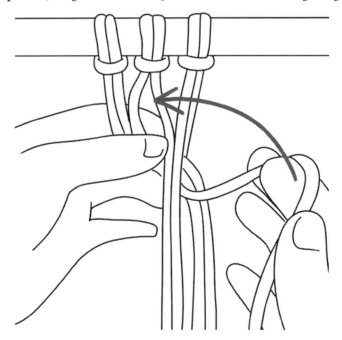

STEP 8: *Continue repeating this pattern.*

When making this knot, just make sure that all the loops are equal for uniformity and elegance. Also, when tightening the knot please remember not to tighten it too much – it should only be pulled together until it's securely firm.

2. Alternating Half Knot Pattern

This pattern can also be made with a contrasting color cord to add variety and vibrancy to the project.

STEP 1: *Start with 3 cords attached to your dowel rod with lark's head knots.*

STEP 2: *Take the 4 cords on the left and make the first half of a square knot (if you don't remember how to do it, please refer to the instructions in the first book).*

Instead of finishing the square knot with the second half, take the 4 cords on the right and make the second half of the knot with them.

Tighten it just enough, without pulling too hard.

STEP 3: *Go back to the left side and make the first half of a square knot with the first 4 cords on the left.*

STEP 4: *Make the second half with the 4 cords on the right.*

STEP 5: *Keep repeating this pattern as desired.*

STEP 6: *Try changing the color of the middle cord to add a special touch if you want.*

3. Square Knot Picot

The square knot picot is a nice decorative element that is sometimes referred to as a "butterfly knot".

STEP 1: *Start with 2 cords attached to your dowel rod with lark's head knots.*

STEP 2: *Make a simple square knot.*

STEP 3: *For the next square knot, tie the first half but leave some space (about an inch) from the previous one, then tie the second half and squeeze to form another square knot.*

STEP 4: *Make the third square knot but leave twice the space you left between the first and second.*

STEP 5: *Make the fourth square knot but leave even more space than between the second and third.*

STEP 6: *Make the fifth square knot and leave the same space you left between the second and third knots.*

Step 7: *For the last knot in the sequence, you will need to leave the same space you left between the first and second knots (1 inch).*

STEP 8: *Take the cords and push all the knots from the bottom up.*

You'll see a beautiful motif appear – that's called a picot and looks like a butterfly, hence the name!

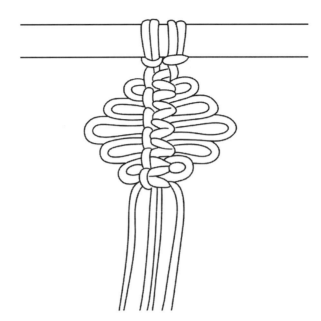

You can repeat the pattern at will. Remember that the more space you leave between your knots, the bigger the picot you will see forming.

3 Original Alternatives to the Lark's Head Knot

One annoying thing about the lark's head knot is that it can be unstable and loosen easily. So here are some alternatives to the lark's head knot.

1.   Reinforced lark's head knot

STEP 1: *Take the cord, fold it in half and simply place it over the dowel rod.*

STEP 2: *Take the ends of the cord and move them upwards behind the dowel to form a loop.*

STEP 3: *Thread the two ends into the loop and tighten.*

The result is a reinforced lark's head knot with an elegant diagonal bar.

The diagonal can be either to the right or to the left.

2. Slingstone Hitch

STEP 1: *Take the cord, fold it in half and pass the "loop" over the dowel rod.*

STEP 2: *Instead of threading the cords through the loop as in the case of a simple lark's head knot, wrap each of the cord ends through the side of the loop that has formed.*

STEP 3: *Pull to tighten the knot.*

You can also do this with an inverted lark's head knot.

3. Cat's Paw Knot

STEP 1: *Take the cord, fold it in half, and flip it over the dowel rod.*

STEP 2: *Now thread the cord ends as you normally would with a lark's head knot. Don't tighten.*

STEP 3: *Pass the cords through the loop two more times.*

STEP 4: *Tighten and push everything upwards.*

Chapter 3
LEARN 9 ADVANCED MACRAMÉ KNOTS

In the first book, you learned the 9 main basic knots that enable you to independently get creative with macramé. This book aims to take your skills to the next level.

Earlier, you discovered 6 original variations of the basic knots.

Now it's time to give you a detailed explanation of the top 9 advanced knots.

Once you have mastered all these techniques, you will have reached a comprehensive and advanced skill level in this wonderful art and be both capable of creating your patterns, and effortlessly carrying out as many patterns as you want without the need for a thousand tutorials.

Without further ado, let's get started!

KNOT #1 – Square Button Knot

This first knot is also known as Rose Knot. Follow these simple steps to create it.

STEP 1: *Take 4 cords, bend them in half and attach them to a dowel rod with lark's head knots (or the variants thereof that you have just learned). You will end up with 8 cords.*

STEP 2: *Proceed to tie some square knots with the first 4 cords on the left.*

Make at least 4 or 5 square knots.

STEP 3: *Collect the filling cords (2nd and 3rd cords from the left) and thread them through the space just above the knot to form a nice round button.*

STEP 4: *Take the working cords (1st and 4th cords from the left) and make a square knot just below the button you just created.*

STEP 5: *Your first square button knot or rose knot is ready!*

KNOT #2 – Josephine Knot

STEP 1: *Take two cords, fold them in half (you will end up with 4 cords), and attach them to a dowel rod with a lark's head knot.*

STEP 2: *Take the two right cords together (as if they were a single cord) and bend them over themselves to form a D-shaped loop that you will place over the left cords so that they fall in the middle of the loop.*

STEP 3: *Take the end of the left cord and thread it over the loop in the space between the first and second cords.*

STEP 4: *Now insert the cord ends into the loop to the right of the left cord (which, let me remind you, is located under the middle of the loop).*

STEP 5: *Thread the ends upwards again into the space between the cord and the curve of the "D".*

It's easier to do than it sounds. All you must do is bring the ends of the left cord up, then down, and then up again from left to right (clockwise).

STEP 6: *Tighten everything neatly and stretch the knot evenly until you have a kind of sideways figure of 8.*

KNOT #3 – Crown Knot

Also known as the Chinese Crown Knot because it appears to have been imported from Asia. Follow these steps to make it happen.

STEP 1: *For this knot, you will need a cord that you will tie to a dowel rod with a lark's head knot.*

STEP 2: *Take the left cord and wrap it around the cord to the right. Have it form a double loop, still underneath the right cord.*

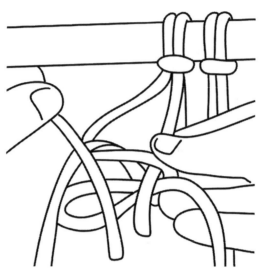

STEP 3: *Take the cord to the right and bring it upwards. Pass it over the double loop and thread its end into the double loop.*

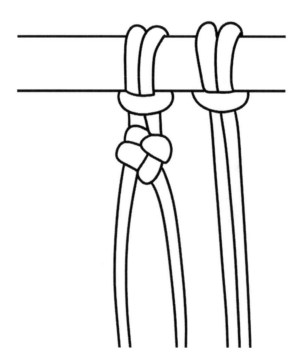

STEP 4: *Now, just tighten the knot carefully.*

KNOT #4 – Pipa Knot

STEP 1: *Take a cord and form a sideways figure of eight. Make a smaller loop to the left and a larger loop to the right.*

STEP 2: *Take the longest cord and pass it behind the top loop.*

STEP 3: *Make a counterclockwise loop placed inside the previous lower loop.*

STEP 4: *Repeat these steps until you can no longer fill the bottom (largest) loop.*

STEP 5: *Now thread the end of the cord through the hole that formed in the lower loop. Your knot is ready.*

NODE 5# Chinese Good Luck Knot

STEP 1: *Take a cord and fold it in half as if you were to pass it over a dowel.*

STEP 2: *Now further fold in half one of the resulting cords to form a narrow loop.*

STEP 3: *Do the same with the other cord. You should end up with 3 loops total (one is the main loop at the head of the cord, and the other two are the side ones).*

STEP 4: *Pick up the free ends of the cord and fold them over the nearest loop counterclockwise.*

STEP 5: *Take this latter loop and fold it onto the next one.*

STEP 6: *Take the last loop and fold it onto the next one. However, this will go under the final loop (the first one you folded).*

Step 7: *Join all the cords and tighten the knot.*

STEP 8: *Repeat all the steps as before. Take the ends of the cord and pass them onto the next loop – and so on, until you return to the first one.*

STEP 9: *Pull all the cords again.*

STEP 10: *Now all you must do is pull the corner cords to create 4 small even loops. And here's your first Chinese Good Luck Knot ready!*

KNOT #6 – Monkey Fist Knot

STEP 1: *Take a long cord, bend its end and make a knot that will serve as a stopper. Just make a small round knot and cut the end of the remaining cord.*

STEP 2: *Hold the cord between your index and middle fingers (with the small stopper knot under your two fingers) and wrap the cord around your index fingers first, then make two rounds around your index and middle fingers together.*

STEP 3: *Wrap the cord around your index finger again, then wrap the cord three times perpendicularly through the cords wrapped around your fingers.*

STEP 4: *Remove everything from your fingers, then thread the end of the cord through one of the loops.*

STEP 5: *Wrap the cord three times horizontally through the other loop.*

STEP 6: *To finish the knot, find the starting point and start pulling the cord. Neatly work on each wrap in sequence and keep tightening the cord until you reach the end. The result will be a spherical knot, like a ball, that you can use as a keychain.*

KNOT #7 – Triangle Knot

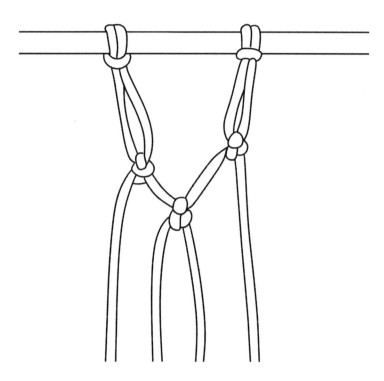

STEP 1: *Take two cords and attach them to a dowel rod with the lark's head knot (or one of its variations if you like).*

STEP 2: *Take the leftmost cord and form an open loop facing right.*

STEP 3: *Take the other end of the same cord and place it over the loop.*

STEP 4: *Thread the end of the right cord into the loop you created on the left. Pass the end of the cord over itself and again onto the left loop.*

Tighten to form a nice triangular knot.

STEP 5: *Do the same with the pair of cords on the right.*

STEP 6: *Repeat the knot at a distance below with the middle cords, to form a nice triangle.*

KNOT #8 – Barrel Knot

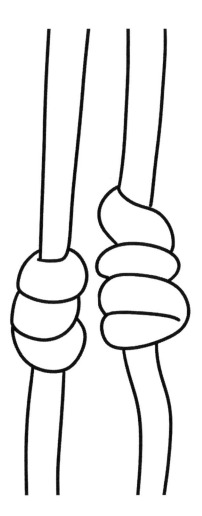

STEP 1: *Take two cords and attach them to a dowel with the lark's head knot (or one of its variants) about 5 cm (1.2") apart.*

STEP 2: *Take the right cord of the left pair of cords and bend it on itself to form a loop.*

STEP 3: *Thread the end of the cord through this loop several (3-4) times. Finally, push the cord upwards by pulling it slightly.*

KNOT #9 – Endless Falls Knot

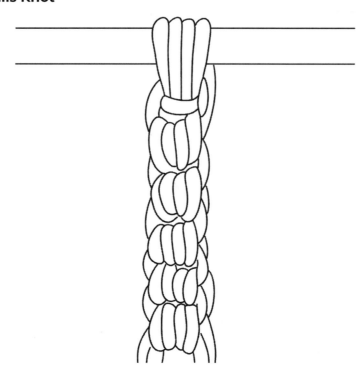

STEP 1: *Take a cord and attach it to a dowel rod with a lark's head knot. But don't tighten the knot, leave it loose.*

STEP 2: *Take a second cord and thread it through the loop you created by leaving the lark's head knot loose.*

STEP 3: *Tighten the knot and fasten the cords tightly to the dowel.*

STEP 4: *You will end up with two central "filler" cords and two external "working" cords. Take the working cords and cross them to form a loop.*

STEP 5: *Take the right filler cord and pass it over the intersection and through the loop. Then take the left filler cord and pass it over the intersection and through the loop.*

STEP 6: *Pull the working cords until a knot is formed.*

Step 7: *Repeat all the above steps until the knot sequence reaches the desired length.*

Chapter 4
ROUND OFF YOUR TRAINING WITH
6 ADVANCED TECHNIQUES

1. <u>Wave pattern variation #1</u>

With this technique, you will be able to create a wave pattern using a square knot weave.

STEP 1: Cut 5 cords, fold them in half, and attach them to a dowel rod with lark's head knots (or any of the variants you just learned).

STEP 2: Make a square knot right in the middle. Use the 4 middle cords to do this.

To make the square knot, start from the right cord, slide it on top of the two in the middle, and behind the cord on the left. Finally, the left cord must be threaded into the loop that has formed. Tighten everything together to obtain a square knot.

STEP 3: Take the left cord and make another square knot on the opposite side.

STEP 4: Take the two working cords you were using to make the square knots and pull them out of the way (lift them over the rod so they don't get in your way).

STEP 5: Take the 4 remaining center cords and make another square knot starting from the right cord.

STEP 6: Again, take the working cords and lift them over the rod so they don't get in your way.

Step 7: Now comes the first part of the wave. Take the leftmost cord and pass it over the first cord you set aside (left side) and under the second one you set aside.

STEP 8: Do the same on the right-hand side.

STEP 9: Use the 4 cords you now have at the forefront to make another square knot around the 2 middle cords. Start from the right.

Make sure the cords are all nice and tight but don't overtighten them.

STEP 10: Make another square knot, this time starting with the cord on the left-hand side.

STEP 11: Take the two working cords in your hand and pull them over the rod as before.

STEP 12: Take the leftmost cord and pass it first underneath, then over, and then back under the 3 cords you pulled over the rod.

STEP 13: Do the same with the right cord.

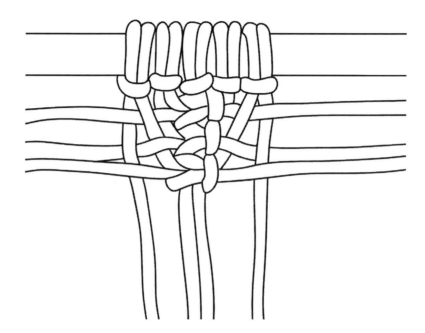

STEP 14: Make a square knot, again starting with the right cord.

STEP 15: Make another square knot starting with the left cord.

STEP 16: Take the two working cords and lift them back over the rod.

The top section of the pattern is completed. Now follow the steps for the lower section.

STEP 17: Take the first cord of the 4 that you had set aside and weave them through the other 3. First pass it underneath, then on top, and then underneath again. Do the same both left and right.

STEP 18: You will now find yourself with 4 cords again. Proceed to tie a square knot starting from the right.

STEP 19: Arrange the working cords as you did with the others.

STEP 20: Take the next cord between the ones you placed over the rod (starting left) and pass it over, under, over, and again under the other three.

STEP 21: With these 4 cords, make a square knot starting from the right.

STEP 22: Take the outermost working cords again and drape them over the rod.

STEP 23: Take the draped cords from the top and pass them under, above, and under the other 3 (proceed both left and right).

STEP 24: Tie another square knot starting with the right-hand cord.

STEP 25: Again, drape the outer cords over the rod.

STEP 26: Take the cord draped at the top and slide it over, under, and then back over the other 3 cords. Do this both left and right.

STEP 27: Tie the last square knot, starting from the right-hand side.

STEP 28: Let all the draped cords drop down and adjust the pattern by pulling where needed to give it a uniform and clean look.

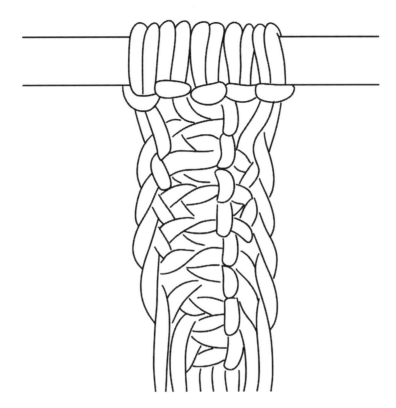

This weave is incredibly versatile. You can use it as a pattern for wall hangers, plant hangers, belts, hammocks, and virtually anything else you can think of.

The more cords you add, the bigger the pattern gets.

2.   Wave pattern variation #2

STEP 1: Attach 3 cords (folded in half) to a dowel with the lark's head knots.

STEP 2: Take the far-right cord and bring it diagonally to the left over the other cords.

STEP 3: With the other cords, tie 5 DHHK knots diagonally around this cord.

STEP 4: Take the two cords on the right and make 7 alternating half knots.

STEP 5: Take the next two cords on the left of the ones you have just worked with and make 4 alternating half knots as before.

STEP 6: Take the leftmost cord and fold it down diagonally to the right.

Step 7: Tie 5 DHHK knots around this cord.

STEP 8: Take the two cords on the left and make 7 alternating half knots.

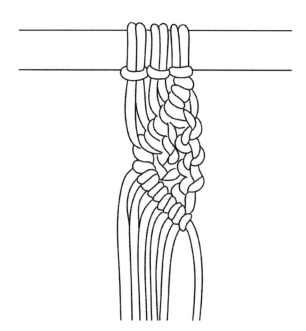

STEP 9: Take the other two cords to the left of the latter and tie 4 alternating half knots.

STEP 10: With the rightmost cord, make another diagonal to the bottom left and tie 5 more DHHK knots.

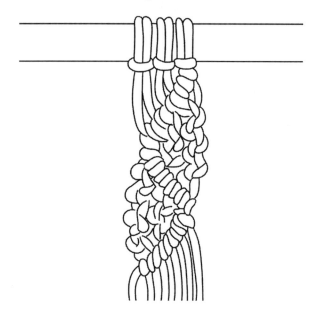

As you can see, you are creating a beautiful pattern with a wave shape. You can repeat the pattern as many times as you want or need.

This pattern is also ideal for wall hangers, plant hangers, and many other projects.

3. Create your own net shopper bag

Square knots can also be used to create a mesh weave for bags, plant hangers, or larger pieces.

Make sure to use a cord made of durable material. The cord thickness should be medium (please refer to the thickness chart earlier in the book). It always depends on what you wish to make. Just make sure that it's not too thin, or it won't be sturdy enough for your groceries.

By using a separator, you will make sure that the knots are evenly spaced.

Making a shopper bag with a mesh pattern

Here are step-by-step directions on how to make your bag. You can use this method also to easily make fabulous plant hangers.

STEP 1: *Take a cord holder.*

STEP 2: *Cut 32 pieces of cord, each about 3m (about 120 inches) long.*

STEP 3: *Fold the cords in half and attach them to the dowel.*

STEP 4: *Grab the first 4 cords and tie a square knot.*

STEP 5: *Repeat 8 times.*

STEP 6: *Pull the cords so that the 8 knots "slide" to the back of the dowel.*

Step 7: *Move to the side so that you face the profile of the rod and take the two cords at the front and the two at the very back to create the side of the bag.*

STEP 8: *Tie a square knot with them.*

STEP 9: *Move back to the front and create the first row of 8 square knots.*

STEP 10: *Use a cord separator to ensure the space you leave between the knots is even.*

STEP 11: *Make the second row of 8 square knots, all at the same distance.*

STEP 12: *Remove the separator and make a row of 8 square knots at the back.*

STEP 13: *Place the separator back on this side and use it to tie another row of 8 square knots.*

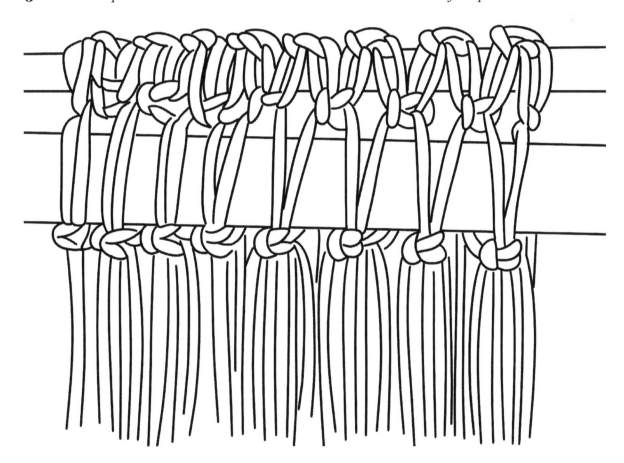

STEP 14: *Repeat this operation up to about half the length of the cords (the number of rows will depend on the space you leave between one square knot and the other). Keep using the separator for reference.*

Now it's time to close your bag with handles.

STEP 15: *Group the cords into 4 bundles, 2 at the front and 2 at the back.*

STEP 16: *Twist the cords of the two front bundles upon themselves. Help yourself with a piece of duct tape to fasten everything well.*

STEP 17: *Repeat this with the cord bundles at the back.*

STEP 18: *If there is any, cut any excess cord from the handles thus created.*

STEP 19: *Finish the handles with 5 meters (about 197 inches) of cord that you will roll to wrap the length of the handle (wrapping technique).*

4. **DIY Macramé Feathers: 3 ways to make them**

Macramé feathers (or leaves) are a very fashionable decorative element suitable for many macramé projects such as earrings, key rings, tapestries, and more.

Here are 3 of the best techniques to easily create elegant macramé feathers. But before we get started, let me give you 4 useful tips:

1. *Do you have any leftover cord from past projects?*

Most likely, you will have many leftover pieces of cord from other projects, possibly of different colors too. The best way to repurpose them is to make decorative feathers.

2. *Making macramé feathers doesn't call for any special kind of cord – actually, you can make them with any type and color, as long as the cord material is a natural fiber, and the cord is woven or wrapped.*
3. *All the different techniques for making leaves/feathers are very similar to those for making fringes.*
4. *If you want to make very large macramé feathers, it might help to put felt on the back of the leaf so that the fringe stays in place.*

Macramé feather: method #1

sTEP 1: *Grab some leftover cotton cord (or whatever you have at hand) to use as a center cord. It should be about 50 cm (about 20 inches) long.*

STEP 2: *Also get other scrap pieces of about 28 cm (11 inches) each.*

STEP 3: *Take one of your shortest pieces of cord, fold it in half, and put it under the center cord a few inches from the top, with the loop facing left.*

STEP 4: *Tie a lark's head knot.*

STEP 5: *Repeat this with another one of the shorter cords but in the opposite direction this time.*

STEP 6: *Keep tying knots on either side with the other cords until your feather is the length you want.*

Step 7: *Now all you need to do is brush your cords. Just use a fringe brush or pet brush and you're done!*

Macramé feather: method #2

sTEP 1: *Again, grab a longer (about 50 cm / 20 inches) piece of cord that you will use as a center cord. Also, get other scrap pieces of about 28 cm (11 inches).*

STEP 2: *Fold the center cord in half.*

STEP 3: *Fold a piece of cord in half and place it under the long center cord a few inches from the top of the loop.*

STEP 4: *Take another shorter piece, fold it in half, and put it on top of the first cord and of the center cord, with the loop end facing in the opposite direction.*

STEP 5: *Thread the ends of each cord through the loop of the opposite cord. Pull to tighten around the center cord.*

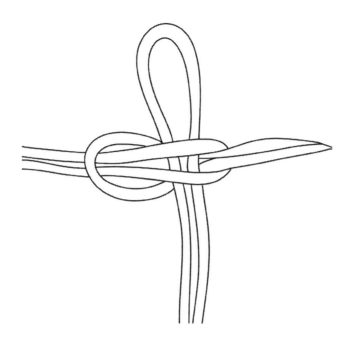

STEP 6: Repeat the same operation but from the other side.

Step 7: Continue to knot the cords in this way around the center cord, alternating sides, until the feather reaches your desired size.

STEP 8: Brush all the cords to form your feather.

Macramé feather: method #3

this technique is slightly more complex. To make the third kind of feather, you will need:

- 1 wooden hoop with a diameter of 2 inches (about 5 cm)

- 3 cords measuring approx. 112 cm (44 inches) each

- 16 cords measuring approx. 25 cm (10 inches) each

- 1 macramé board or cork board

- T-needles (see book 1).

STEP 1: *Take the wooden hoop and the 3 longer cords.*

STEP 2: *Fold the cords in half and fasten them to the hoop with lark's head knots. Secure everything onto the board with pins.*

STEP 3: *Take 2 of the smaller pieces of cord and fasten them with a lark's head knot to the top of the cord on the far right.*

STEP 4: *Repeat this with the cord to the far left.*

STEP 5: *Now, tie two DHHKs on the left side, heading down and to the left.*

STEP 6: *Repeat on the right.*

Step 7: *Tie two more DHHKs heading down and to the right.*

STEP 8: *Tie two more DHHK knots, this time on the right side, heading down and to the left.*

STEP 9: *Merge the two sides with a DHHK heading down and left.*

STEP 10: *Tie 3 more of the shorter cords to the right and left cords with lark's head knots.*

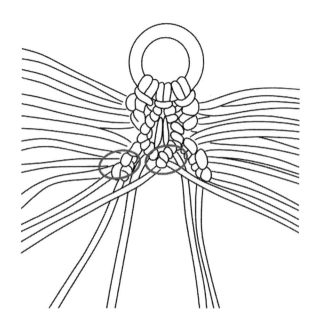

STEP 11: Repeat the DHHKs and continue to add the remaining shorter cords on both sides until you create 3 diamond patterns.

Step 12: All you need to do is brush your cords.

If you end up with different cord lengths when finishing the feathers made with any of these 3 methods, trim down the excess cord for a uniform look.

Chapter 5

HOW TO CREATE YOUR PATTERN

By now, you have learned the top 9 basic knots and the top 9 advanced knots; you have mastered advanced macramé techniques and completed a few different projects.

Now you will learn how to put together this knowledge into a diagram or pattern.

WHAT IS A PATTERN?

what you've learned so far is that all patterns are recurring designs that turn into decorative motifs.

As you have seen, through patterns you can create an infinite range of home decor items and many other objects.

LEARN HOW TO READ PATTERNS WITH A PRACTICAL EXAMPLE

before creating your patterns based on my suggestions, you may want to familiarize yourself with ready-made patterns.

This paragraph aims to make you understand how to read a macramé pattern.

To do this, we shall take a practical example: the pattern of a macramé bracelet.

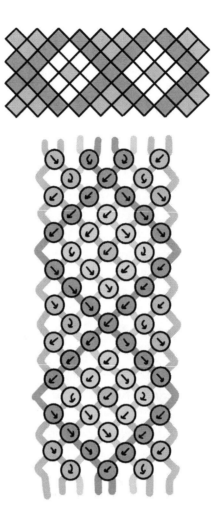

1. *The top drawing shows how the diagram should turn out. The lower part represents the instructions.*
2. *The colored lines are the cords (sometimes indicated with letters of the alphabet), while the circles represent the knots.*

From this diagram, we can see that there is a total of 4 pink cords, 2 white cords, and 2 red cords.

To make things easier, you may want to use a macramé board.

3. *Now, the way to read the pattern is from left to right and from top to bottom.*

So, first, you must create the first row of knots. Then the next one just below, again starting from the left.

The arrows indicate the direction of the knots.

4. *A left-to-right arrow means that you are supposed to tie a double knot with the left cord.*

In this case, the left-hand cord will pass over the right-hand cord, forming a loop on the left. Then it will go under the right cord to slide into the loop.

The operation is to be repeated twice.

5. *A right-to-left arrow means you must tie a double knot with the right-hand cord.*

In this case, the cord will pass over the one on the left, forming a loop to its right.

Then it will slide under the left cord and slip into the loop to form a knot.

The operation is to be repeated twice.

6. *An arrow that goes left to right and then right to left means you must tie a knot and then go back with the cord to the left.*

In this case, the left-hand cord will pass over the right-hand cord, forming a loop on the left.

Then it will go under the right-hand cord to slide into the loop.

The second knot is made in the same way – but in the opposite direction.

7. *An arrow that runs from right to left, then from left to right, means that you need to make a knot and then place it with the cord to the right.*

This cord will pass over the one on the left, forming a loop to its right.

Then it will go under the thread on the left to end up in the loop.

The second knot is made in the same way – but in the other direction.

HOW TO CREATE YOUR OWN PATTERNS

there is a very special technique that allows you to create your own patterns, and it is known as pixel macramé.

The following instructions will enable you to create your own patterns without the need to buy them or try a thousand tutorials.

For example, if you're thinking of creating an original pattern for a wall hanging, you may follow these steps:

The first step is planning. Answer the following questions:

1. *What cord do I want to use? How much cord do I need?*
2. *What kind of design do I want to accomplish?*
1. *Regarding the first question, both in the first and second books we talked about the main 3 types of cord.*

Regarding quantity, remember that the thicker the cord, the less you will need to fill the pattern, but it all depends on the effect you aim to achieve.

2. *Now let's move on to designing your pattern: how do you get started? Follow these steps:*
1. *When it comes to designing, one of the best ways to get started is to sketch what you have in mind to achieve.*

Don't worry, it doesn't have to be precise – it's just a way to visualize what's on your mind by putting it on paper.

My advice is to simplify the shapes as much as possible.

2. *The next step is to arm yourself with a tool that enables you to recreate your drawing in the form of pixels.*

There are many on the Internet; one could be

https://www.pixilart.com/

It's very simple and intuitive.

3. *You could recreate your drawing by starting to select a size of 13x20 pixels.*

The first digit represents a vertical line. In turn, each vertical line will represent a cord folded in half and attached to a dowel or other support.

Since I decided that my wall hanging should be a rectangle, 20 stands for the height (which will be greater than the width).

A function within the program allows you to see each pixel, thanks to a grid.

Now, by following the instructions, you can easily recreate your sketch.

As you can see, the result will be a grid where the individual pixels are clearly visible.

4. *Once you have made your design, you can now also estimate more accurately how much cord it will require.*

For example – according to the pattern – you will have 13 cords horizontally.

Each pixel corresponds to a knot. If you estimate how much space each knot will take up (based on the thickness of your chosen cord) you can easily estimate how much cord you need.

Remember to err on the plus side, because you might need some extra cord for the finishing touches.

Thanks to the pattern, when there are multiple colors you can calculate the number of knots with each different color per row.

Chapter 6

TOP 5 + 1 ADVANCED HACKS

1.   How to keep the fringe straight

If you are doing a project that involves a fringe, but it has not come out well – if it curls or does not stay straight – then here is the solution to your woes.

First, remember to "comb" the fringe well.

Start at the bottom and work your way up, just as you would if you were unraveling knots from your hair.

Using a pet brush is the quickest way to straighten a macramé fringe.

The only potential drawback is that combing the fringe with a pet brush may produce a considerable amount of lint. If you don't want it to stick to your clothes, protect yourself with an apron or a towel.

Once you've managed to straighten your macramé fringe, here's how to keep it straight.

Add a backing: felt or other stiff fabric can be applied on the back with fabric glue to keep your fringe or feathers perfectly straight.

Another trick is to spray some ironing starch. After spraying the starch, give it a quick pass with the iron to dry it and compact everything well.

2. 3 ways to cut the fringe perfectly straight

You are looking at the project on which you have spent so much of your time and energy. You would be happy if it wasn't for… that crooked fringe!

Sometimes, cutting the fringe can be a real hassle: one thread is longer, one thread is shorter… it may ruin the overall effect of all your hard work.

No need to fright, here are 3 methods to cut a perfectly straight fringe.

METHOD #1

* Use adhesive tape. Apply it gently (without fixing it hard) onto the part of the project you need to cut.
* Simply trim along the tape edge.

Remember that the tape must be easily removed. Masking tape works best because its adhesive isn't too aggressive.

If you only have some super sticky tape, first stick it to your jeans or another surface so that it unloads most of its glue.

METHOD #2

* Decide on the length of your fringe.
* Measure each cord with a measuring tape or ruler.

This is a time-consuming but very effective method.

METHOD #3

* Tie a reference cord at the height of the cut you want to make.
* You can also use a laser level to make sure the cord is perfectly straight.
* Simply trim the fringe following the reference cord.

3. What to do when your cord is too short

One of the worst accidents that can happen is running out of cord while you are working on your knots or putting the finishing touches on your project.

Here are some solutions to add more cord to your project and save your hard work!

1.   *You are tying a square knot and you have no cord left (you only have 2 short ends left).*

Take 2 new cords and fasten them with a pin on the short cord stumps.

Finish your square knot.

Remove the pins from the cords and secure the extension cords to the short cords with a knot.

Hide the knot behind or under your project.

After attaching the new cords, just keep working.

2.   *You're tying a DHHK and you're out of cord.*

Again, pin a new longer piece of cord to the short one you are left with.

Keep working with the added cord.

Simply hide the head of the cord you added inside the project.

3.   *Again, you are tying a square knot, but you're running out of your center cords.*

Take a new piece of cord and wrap it over the two short center cords. Tighten them together to form a single cord.

Finish your square knot with the new cords.

Once you tighten the knot, it'll look a little warped. Adjust all the cords until the mess is concealed. If necessary, cut the stumps of the short cord if they are visible.

Proceed with your pattern.

4. How to care for your macramé

Did you stain your macramé?

Don't worry, it can happen. What's important is to find a fix.

If you have just dropped liquid on your macramé, hold the stained section under tap water and let it run for about 5 minutes.

While the water flows, rub the knotwork very gently.

You can also use detergents if they do not contain bleach.

Let dry.

If you can't use running water from the sink, all you must do is dab your macramé with a damp cloth dipped in a mild detergent.

Finally, allow it to dry.

Choose a dry and ventilated place without direct sunlight.

How to keep your project clean

Smaller items should be washed at least every two weeks. For larger items (e.g., curtains, wall hangers, plant hangers) once a month is sufficient.

The cleaning method varies depending on the item.

As a rule, you can start by removing any dust with a soft-bristled brush.

If possible, machine-wash with a mild detergent.

If it's a fragile macramé item, wash it by hand using lukewarm water. Rub gently with a cloth before rinsing. Do not wring.

To clean your plant hangers, what you need are two spray bottles.

Fill one with a solution of mild detergent, bicarbonate, and warm water.

The other should contain cold water.

Spray the detergent solution on the plant hanger, then spray it with cold water. Allow to air dry.

Some golden rules:

- If your macramé is made from natural fibers, over time it could be attacked by parasites. Just keep it well dusted.
- Do not use bleach
- Do not vacuum clean
- Do not place it to dry under direct sunlight.

5. Macramé Finishing Techniques

There are countless methods to finish your macramé.

A proper finish is important not only for the look but also to protect your work by preventing the cords from unraveling.

In the specific case of jewelry, the finish often doesn't consist of knots. You would use metal clips, hooks, fasteners, and so on.

In short, the best method to adopt always depends on the project.

Let me show you some of the simplest and fastest to achieve.

Example #1

This first finishing method is simple but effective for many pieces.

- Finish your project by hand by tying a knot at the end of each cord.
- Apply some clear glue to the end of the knot to prevent the cord from unraveling.

Example #2

- Wrap several cords together and make an overhand knot.
- Cut any excess cord ends.

Example #3

- Fray the ends of the cord. We saw how to do this in chapter 5.

Example #4

- Add beads to the end of your project.
- Cut the cords even, then insert a bead and tie a knot at the end of the cord.
- Cut any excess cord and secure the cord end with a dab of clear glue.

6. Bonus: the perfect playlist for working with macramé

We talked about the meditation and mindfulness properties of macramé handiwork in the first book. Now, how about a bonus tip to make every moment you devote to macramé perfect?

Here is a sample playlist of 20 songs that will help you work with even more concentration and relaxation, thanks to some wonderful background music.

Of course, this is just a starting point. As you get better and better at knotting, you will also be able to select your individual macramé playlist.

1. Lean In – Rising Appalachia
2. Siete – Nicola Cruz
3. Synergy – Tash Sultana
4. Cocoon – Milky Chance
5. Kiara – Bonobo
6. Prayer in C – Lilly Wood and The Prick
7. The Fader – Beats Antique, Gaudi
8. Clementine Jam – Orchestra Baobab
9. Wish I Knew You – The Revivalists
10. You're The One – Kaytranada, Syd

11. Clair De Lune – Flight Facilities, Christine Hoberg
12. Harvest Moon – Poolside
13. Anemone – Slenderbodies
14. Taro – Alt-J
15. Monday Loop – Tomppabeats
16. Where Angels Fear To Tread
17. Down For It – Habitaat
18. Clap Your Hands – Whilk&Misky
19. So Young – Portugal The Man
20. Santurce – Twuan

Chapter 7
TURNING YOUR PASSION INTO A BUSINESS

Did you enjoy learning macramé so much that now – with all that you have learned, and you have become so good at it – you would like to earn something from it? Just know that it is possible to do so.

Indeed, there are macramé artists who, thanks to e-commerce and social networks, have been able to start online businesses.

Of course, there is no guarantee that you will be able to fully support yourself with macramé; it much depends on your skills and resourcefulness.

But if you so wish, it can easily become a considerable side hustle.

Keep in mind that the advent of Etsy and YouTube has opened up unprecedented possibilities – and at the same time, caused a lot of competition.

Some people started publishing simple tutorials on YouTube and then went on to become real macramé stars.

WOULD YOU LIKE A SUCCESSFUL BUSINESS? HERE ARE 4 THINGS TO BEGIN WITH

these are the steps you absolutely must follow to have a better chance of success:

1. Give your business a name

Naming your craft business means building something of your own, and it will fill you with pride.

But the most important thing is that it will give you better chances to be found and remembered, right away. That's if you follow some tricks for choosing the right name. Especially if you plan to sell your macramé online, you might need a domain name:

- **It must be unique.** If you want to open your online business, you need a domain for your e-commerce. Therefore, the domain name must not be taken already. Think also of the social pages on Instagram and Facebook, as well as a YouTube channel
- **It must be short and meaningful.** A business name must be straightforward. Try to include any keywords that immediately represent your business
- **It requires a touch of originality.** You should not choose a banal name, or else you risk being confused among the competition.

My suggestion is to create a logo and print business cards with your business name and contact details.

2. Learn how to estimate the right selling price of your items

If you want to look professional, your items must have coherent pricing based on specific factors:

- **Supplies cost.** You need to be able to factor in not just the cost of the raw materials, but also any tools you purchased.
- **Labor (hours).** Decide what your hourly rate should be.
- **Competition.** What do your competitors do? That is, what are the asking prices of both local and online businesses in your niche?
- **Who you are selling to.** Are you selling online, or to family and friends?
- **Shipping costs.** If you sell online, you need to include packaging and shipping expenses.

3. Keep your production costs low

If you are just starting out, you need to keep your costs to a minimum. My suggestion regarding supplies is to opt for low-cost materials and wholesale prices.

Especially if you want to undertake larger projects, you need to be able to estimate the exact cost of all the cords you are going to use.

4. Use social media to promote your business and grow your brand

Social media – Instagram and YouTube especially – are very important to promote your craft business.

With its many groups about macramé, Facebook can also be a great avenue for getting your name out there.

11 WINNING STRATEGIES TO START EARNING WITH MACRAMÉ

you are now ready to start your own business. Here are 11 winning strategies to start earning money right away:

1. Sell your work to friends

Before launching your online presence, practice managing your business by selling the items you create to friends first.

It will help you gain experience, make yourself known, and generate active word-of-mouth.

Inviting your friends to dinner is a great opportunity to show them your projects. Since macramé is very fashionable, don't be afraid to tell them what you do.

Beware: if they ask you for projects that are too challenging for a beginner, don't stretch yourself. Just say no and suggest alternatives.

2. Sell your work at local fairs or craft markets

Remember that as a rule, people who attend these markets are not prepared to spend much. The best items to sell at a market are:

- plant hangers
- small items such as key rings or bookmarks
- shopper bags
- small wall hangings
- handmade jewelry
- Christmas decorations.

3. Introduce yourself to stores in your neighborhood

If you live near boho-style venues or craft stores, pitch yourself to them as well.

Select your best pieces to introduce yourself. They may decide to sell your items or allow you to display them in their spaces.

4. Create your own online presence

Nowadays, creating a website or e-store is very simple. However, get professional help if you can. A freelancer is quite inexpensive, but you need to make sure their skills are proven.

5. Open your YouTube channel

If you're into video making and teaching, tutorials can be your thing.

At first, you can make videos even just with your mobile phone and a tripod. If you continue reading, you will find all my tips to create professional tutorials even if you are not a video maker.

Here are a few tips to get you started:

- do not make long videos (keep them under 10 minutes if possible)
- create video covers with Canva
- use targeted keywords in your video titles
- provide a clear and detailed description
- use consistent tags.

6. Organize face-to-face workshops

As we said before, macramé is experiencing a golden moment. Many people in your neighborhood may want to learn how to do some small macramé projects.

You could set up a workshop to teach them the basics of macramé. It would be a great opportunity to get your name out there.

You could also create a flyer, print a hundred copies, and hand them out in your neighborhood. No need to be a graphic designer, there are intuitive apps like Canva that are very useful for this.

Or you can advertise it on your social pages or in local Facebook groups on macramé.

7. Create an online course

Don't be scared! You don't need to create your course from scratch. Just sign up on a dedicated platform, record your videos, and you're done.

8. Start a blog

Starting a blog can be very challenging but it's a great way to make a name for yourself and sell your products.

Also, if you have enough traffic, you can make money through ads or affiliate marketing.

9. Sell through online platforms like Etsy

There is a lot of competition on platforms like Etsy or Shopify, but if you follow these tips, it will be easier to stand out from the mass:

- strive to create unique and surprising items
- use fashionable colors
- take photos with adequate lighting and, if necessary, edit them with photo editing apps to make them look more professional
- write compelling and detailed descriptions.

Before opening your store on these platforms, make sure to check their fees, since they are yet another cost that you need to factor into your pricing.

10. Sell your patterns online

You can choose whether to sell online your actual items or the designs you followed to make them (provided they were created by you).

The upside is that earning money would require less energy and time, plus it would significantly strengthen your brand.

Videos are more convenient for people, but you can also use patterns with illustrative photos.

Before selling them, have some people try them. They can provide valuable tips and help you fix any mistakes.

Before selling your designs, you may also consider purchasing some professional ones to study them and understand how they work.

11. Sell macramé cords online

Many crafters and artists have started online businesses as craft suppliers. People trust them to buy supplies from because they give honest opinions about cords.

Of course, the margin you can get from a spool of cord can be low, so you need to evaluate if it's worth your time and if you can generate a sufficient volume of demand.

STARTING A CRAFT BUSINESS: THE MAIN PRACTICAL ASPECTS

i am not trying to discourage you, but you need to keep in mind that starting a craft business is a full-time job.

There are so many activities you might not have considered:

- opening and managing an e-store

- managing your social pages and online marketing

- managing your stock and supplies
- ensuring you have enough space to store the materials
- photographing the items and creating the copy for product descriptions
- packing and shipping the items as well as handling any returns
- being available for customer support 7 days a week and knowing how to handle any complaints
- handling the financial aspects of your business.

BONUS #1: HOW TO SELL YOUR MACRAMÉ ON INSTAGRAM

first of all: why Instagram?

- Instagram has evolved to be one of the most popular social apps in the world. Just consider that it has about 1 billion active users.

With such figures, you can easily understand that the audience of potential buyers – even if you work in a niche like macramé – is still huge.

- It is very easy to use, and it showcases top-quality content. Its interface is very intuitive and highly responsive.
- It uses hashtags. I'm sure you know that a hashtag is a word preceded by the # symbol. Why are hashtags so important and popular? Many use them quite randomly, but as a matter of fact, they have a very specific function: to immediately identify the topics that interest you.

For example: Are you looking for macramé patterns? Just type #macramépatterns to find all the content you want.

- You can easily apply filters and effects to your photos, add texts, and publish short videos… In other words, you can get creative and achieve professional results even if you are anything but an expert.
- And best of all, this is free. Of course, if you want to publish ads those are a paid feature, but if you just want to download and use the app, you can do it for free.

But let's get to the most important part…

What can Instagram do for your macramé business?

Instagram is a fantastic online showcase. You can publish as much content as you want and promote your business virtually at no cost.

By creating a business profile, you can enter all your data, a description of what you do, receive messages, and even add a business email if you have one.

You can also access all the insights regarding your page to learn more about your target audience or see metrics about your followers, likes, and comments.

But there's even more to it! With the shop function, you can connect your online store to this social network.

The result is that in each of your photos you will be able to insert a link that will take Instagram users directly to your store where they can make a purchase.

With a platform with so many users, you can gain huge visibility and leverage it to do business. Thanks to Instagram, people will get to know you and appreciate what your business has to offer. Consequently, they will be more likely to buy from you.

6 tips to make the most of Instagram for your business

1. **Use appropriate hashtags.** Hashtags should not be chosen randomly but should be consistent with your business and what you are publishing in your posts. Here are some examples:

#macrame

#macrameartist

#macramewallhanging

#weaversofinstagram

#handmade

#bohostyle

#macramemakers

#macramemovement

#macramecommunity

By adding a hashtag, you give yourself a huge opportunity to reach people looking for something related to macramé and attract them to what you do!

2.   **Publish relevant content, frequently.** Don't get out of bed in the morning thinking, "I'm going to post this on Instagram." That's something you do on your personal profile. You need to make sure you publish posts that are consistent with what people are looking for to keep them engaged.

Of course, that doesn't mean you should be communicating formally – indeed, far from it!

3.   **Use your Instagram stories to speak directly to your audience.** Stories should be used either to establish a direct relationship or to publish exclusive content on the "behind the scenes" of your business. They are visible for 24 hours, so you can publish flash content without cluttering your main page.

4.   Schedule your posts. Scheduling your posts helps you create appealing captions and the kind of high-quality images that are appreciated on this social media platform.

5.   Keep an eye on your competitors. Don't underestimate the power of understanding what others are doing to improve yourself and find inspiration.

6.   Collaborate with other accounts. You can also decide to network with other activities or people (influencers) who might be interested in what you do. Remember that social media is first and foremost about networking.

7.   Expand your audience by linking your Facebook page to your Instagram profile. This will give you even more visibility.

How to create a business profile on Instagram

Creating a business profile is very simple. Just follow a few quick steps:

1.   Download the App
2.   Create an account (you may want to use your business email)
3.   Create a username (ideally, it should be related to your business name)
4.   Fill your profile with a description and contact details.
5.   Upload a profile picture (e.g., your business logo)
6.   Start posting.

But first, think through your overall marketing strategy by answering the following questions:

•   who is your audience?
•   What age group are they?
•   What are you planning to sell?
•   What kind of content do you want to publish?

If you want to make the most of what Instagram has to offer, you need to set out clear goals and the step-by-step strategy you plan to implement. Remember to check your page insights to understand the feedback you receive and what kind of posts solicit the most interactions.

Bonus #2: A complete guide to creating quality YouTube tutorials without being an expert

This section of the book is a comprehensive guide that will show you how to produce high-quality YouTube tutorials without professional equipment.

To begin with, tutorials are for the most part explanatory videos made with voiceovers.

YouTube has a built-in feature to produce subtitles and international translations in most world languages.

The duration of your video tutorials should not exceed 10 or 15 minutes, because people are intimidated if they see that a video is too long. Time is valuable for everyone, and there we all have a thousand things to do.

STEP 1: Prepare for the recording

Before you start recording, a great tip is to create an outline of the video you want to make, so you don't run the risk of getting lost while filming.

Don't make it too wordy but use lists and numbered items instead. You must read it several times to get the points you are going to make well embedded in your memory.

STEP 2: Use a recording App

One of the best is Camtasia. It has many useful features and effects to spice up your video.

The only flaw is, that it's not free. It costs $ 299, but you have the option of trying the free 30-day version first.

Remember that to film high-quality videos, you must choose the 1280×720 Full HD option.

STEP 3: Split the recording

While recording it would be best to keep the video separate from the audio, i.e., record the tutorial first and then the voiceover with the instructions.

This minimizes errors and therefore the time required to make any changes.

STEP 4: Film the video

Be careful while recording your video tutorial. If you are teaching how to make a knot, show it step by step and slowly. Try not to make mistakes or unnecessary actions that may be confusing. Keep your movements clear and clean.

Don't be discouraged if, in the beginning, you make more mistakes than you'd like to: it's perfectly normal!

IMPORTANT: Make sure your project is properly lit. You can do this by using two or three light sources.

STEP 5: Record the voiceover

Write a script for your video before recording the voiceover. It's an extra job but it will make you stand out from most of the tutorials found on YouTube.

If you read a script while recording the voiceover, any errors will be greatly reduced.

Record in a quiet, noiseless room and try to keep your voice clear and natural. Do not speak in a dull tone, but make sure to emphasize with your voice the most important passages of the video. That would be the steps that require more concentration.

STEP 6: Main edits to the recorded file

* Size: as previously mentioned, the ideal recording resolution is at least 1280 x 720. This is considered a standard resolution.
* Zoom: you can use the zooming function when showing text or a website URL. The duration of this effect can be set to 1 second for smoother zooming.
* Transition: use transition effects between different segments of the video. There are several to choose from: Page turn, Cube rotate, Fade, Fade through black. You can also set the duration of these.

Step 7: Your video is almost ready

* Export your tutorial. Do not use the Camtasia default options, because otherwise the video will be darker than normal, and the details will be blurred.

Here's how to export a high-quality video:

1.   **Under** *Custom product setting***, click Add/Edit preset. Choose "New" to create the new production preset. Give your setting a name.**

2.   **Choose AVI as the video file format to have high-quality videos with good light.**

3.   **Set the frame rate to 15 frames per second (fps). This enables you to have a high-quality video that weighs much less.**

4.   Create the subtitles file for your video. You can use one of the free apps that you can find on the Internet. Remember that subtitles must be exported in SRT format

STEP 8: Upload your tutorial to YouTube

● Write a title and description

The title of your tutorial is hugely important. It must include keywords that are consistent with the content.

It must be a maximum of 70 characters long. Please note that only the first 40 characters of the title will be displayed when the tutorial is shown as a related video. Therefore, your keywords should be at the beginning.

As for the description, put the most important concepts in the first 3 lines, so that they can be displayed before disappearing under the "Show more" feature.

● Utilize tags, because they are useful to make your video be displayed under the same topic to more viewers in your target audience. Use long tail keywords and include as many tags as possible.
● Make sure to set your privacy settings to public and remember to make your video available for mobile devices and TVs. Enable comments and replies.
● Remind viewers to subscribe to your channel and add links to your sales channels and your other tutorials, useful information, and more.
● Group your videos by topic to form a playlist. This allows users to easily follow all your tutorials without having to search for related videos.

STEP 9: Keep recording!

Now that you've gotten the hang of it and realized that making professional videos is easier than you thought, keep recording to trigger the full potential of your channel!

ADVANCED GLOSSARY

C

Chinese Macramé: knotted designs originally from China and other Asian countries.

Crown Knot: A decorative knot also known as "Chinese Flower" because it resembles a flower.

Combination knot: Using two or more knots together to form a new type of knot or design element.

F

Fusion knots: *Another term for combination knots.*

Folding: *when you are asked to bend one cord around another.*

Forward: *when the cord is heading towards the front of the project board.*

G

Gusset: *this term indicates the sides of a 3D project, such as a bag.*

M

Mat: a kind of knot that is completely or partially filled in when tightened. It is usually made with two or more cords.

Metallic Yarn: a type of cord treated to look like silver, gold, or copper.

Micro-Macramé: macramé projects that are made with very fine cords. Micro-macramé is often defined as any macramé that uses cords thinner than 2 mm in diameter.

Mounting (or Mount): the early step of a macramé project, i.e., when you attach the cords to mounts such as rings, dowels, bag handles, jewelry clips, or other cords.

O

Organize: when you divide the cords into groups, before tying certain knots.

P

Picots: Small loops created when making a series of square knots by leaving some space between them. When you push the square knots together, loops form on the side.

S

Symmetry (or Symmetrical): *a design that is equal on either side or at the top and bottom.*

T

Tension (or Taut): *cords secured so that they do not bend or move.*

U

Unravel: *the process of separating the fibers of a cord. You do this when you want to create a fringe or a feather.*

V

Vertical: *Knots or patterns running from top to bottom.*

W

Working Cord: *the cord used to tie the knots.*

Working End: *name given to the end of a cord that does all the movement when forming the knot. As opposed to "standing end".*

Conclusion

You have now reached the end of the second book. Congratulations! This means that you have now learned:

- advanced variants of the basic knots
- advanced knots
- advanced techniques
- how to fuel your creativity by creating original patterns
- how to turn your passion into a business.

It's now time to delve even deeper into the thick of things. How about learning how to make beautiful plant hangers and decorate your home and garden with beautiful handicrafts?

Read on in my next book to see how to create these wonderful additions to adorn any home.

BOOK 3
MACRAMÉ HOME GARDEN & PLANT HANGER

**Remember to check out your Easy Download
and Print Instructions with Extra Pictures
for all the Projects in this Book**

Please scan the QR code or follow the link on Page 166

INTRODUCTION

Macramé Plant Hanger: a Must Have Item

Macramé plant hangers are becoming must-have items for boho decor lovers and beyond. These furnishings create a warm and quirky atmosphere in any environment, as well as being fantastic space-savers.

They are also a real elixir of life for your plants because they allow better drainage and easier access to sunlight – thanks to the plant hangers, plants can be easily moved where needed, including between indoors and outdoors.

As you know, plant hangers are made by tying together textiles and cords, and perhaps embellished with wooden beads, tassels, or some such.

All you have learned so far enables you to make a plant hanger in just one afternoon.

Obviously, the degree of complexity of the pattern also depends on your experience.

Eager to get started? Jump straight to the chapter detailing the best projects to create different types of plant hangers.

What You Need to Make Your Plant Hanger

Here are some things you should have on hand to make a macramé plant hanger.

1. Cords: for plant hangers, the best cords are cotton cords because they are soft, flexible, and easily available. They are also easy to tie and do not stretch over time.

Alternative materials such as hemp, jute, or polypropylene can also be used (the latter being a great option for outdoor plant hangers).

2. Scissors: to cut the cords.

3. Tape measure: to measure the cord lengths.

4. Plant pots: pick the size and shape you like best. Remember that the heavier the pot, the stronger the mount must be (we will talk about it in the next paragraphs).

5. Wooden or metal ring (optional): many plant hangers feature wooden rings of different sizes to attach to the hook.

Alternatively, you can have a hook with an eyelet. More about this later.

6. Hammer and hardware: unless you already have a dedicated spot to hang your planter, you may need a hammer, nails, or other hardware to install suspension points on the walls or ceiling.

8+1 ways to attach and mount the cords

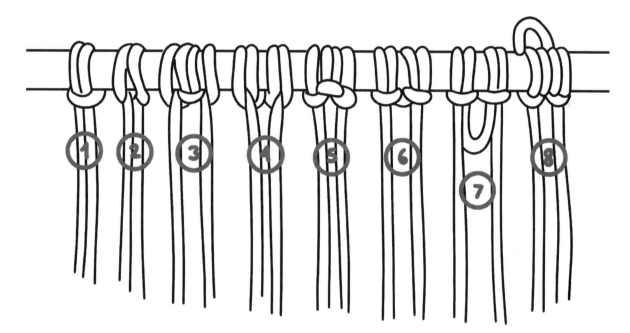

1. **Lark's Head Knot**

This is probably the most common method to attach cords to a stick, dowel, or whatever.

STEP 1: As we saw in the first book, to do this you just need to take your cord and fold it in half.

STEP 2: Slide it halfway over the dowel (or whatever your support is).

STEP 3: You will end up with a loop behind the dowel. No, all you need to do is thread the loose cord ends forward into this loop, and pull tightly.

2. **Reverse Lark's Head Knot**

This method is very similar to the previous one.

STEP 1: Once again, you need to take your cord and fold it in half, except this time you will slide the cord behind the dowel so that the loop sits at the front.

STEP 2: Next, take the two loose ends of the cord and thread them through the loop that was created. Finally, we pull hard to tighten, and the knot thus formed is a Reverse Lark's Head Knot. The only difference is that in this case, the line fastening the cords remains hidden.

3. **Lark's Head with Half Hitch Knots**

STEP 1: This method involves starting with a cord knotted into a lark's head.

STEP 2: Next, you will tie a half hitch knot on each cord.

STEP 3: Starting with the left cord, send it over the dowel or stick to form a loop. Thread the cord through that loop and pull it tight.

STEP 4: Repeat the same process with the right cord.

This method is visually beautiful and allows you to put some space between the two cords.

4. Reverse Lark's Head with Half Hitch Knots

STEP 1: First, tie a lark's head knot in reverse with the cord.

STEP 2: Repeat the steps as above – that is, take the left cord and create a loop over the dowel, through which you will thread the cord.

STEP 3: Do the same with the right cord.

5. Lark's Head with Reverse Half Hitch Knots

STEP 1: Start with a lark's head knot. This time you will tie reverse half hitch knots side by side, both left and right.

STEP 2: Starting with the left cord, create the loop behind the dowel (just bend it into a ring behind the dowel).

STEP 3: Bring it forward from below and thread one of the cords.

STEP 4: Tighten well to knot.

STEP 5: Repeat the same process with the right cord.

6. Reverse Lark's Head with Reverse Half Hitch Knots

STEP 1: First, proceed to tie a reverse lark's head knot.

STEP 2: Next, make a reverse half hitch knot with the left cord following the steps detailed above.

STEP 3: Repeat the same steps with the right cord.

This is a great method to adopt if you want some space between the cord and an excellent visual result. Indeed, the result looks like a single reverse lark's head knot.

7. Reverse Lark's Head with Loop #1

STEP 1: Start with a lark's head knot.

STEP 2: Do not tighten the knot completely, but leave it a little loose instead.

STEP 3: Take the left cord and tie a reverse half hitch knot as before.

STEP 4: Leave this knot a little loose as well.

STEP 5: Repeat the same steps with the cord on the right.

STEP 6: Now, pull down a little the middle (lark's head) knot that you left a little loose.

Step 7: Pull the ends of the left and right cords.

STEP 8: As you can see, the result will be a cute small ring hanging from the center, between the two cords.

This method enables you to quickly add a touch of creativity and originality to your pattern.

8. Reverse Lark's Head with Loop #2

STEP 1: Once again, start with a lark's head knot.

STEP 2: Follow the same method as before, and don't tighten the knot all the way.

STEP 3: As before, tie two reverse half hitch knots both left and right.

STEP 4: Take the hanging part in the middle, and instead of pulling it down pull it back until it reaches the top of the dowel.

STEP 5: Now, pull together all the cord segments on the dowel that you had left a little loose.

STEP 6: Tighten each of the two hanging cords.

Step 7: The result is a pretty loop above the dowel.

This is another great decorative method, not to mention the fact that you can utilize the resulting loop to hang your entire project. It's worth mentioning that this kind of attachment works best with stiffer types of cord.

9.   Cord loop with square knots

There are attachment types for which you do not need dowels or wooden rings. Simply knot the cord to form a loop by which you can hang your plant holder.

STEP 1: Take the cords you want to use to make your plant hanger. Let's say for example that you want six.

STEP 2: Fold the cords in half.

STEP 3: Take another cord and start knotting a series of square knots 2 inches above their middle point (half-point of the cords). The cord with which are tying the knots runs under the batch of 6.

Work on a table or desk to tie the knots more easily.

STEP 4: Fold the set of knots you made to create a cord loop.

STEP 5: Take another cord and secure the loop with a wrap knot.

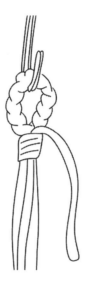

Chapter 2

HOW TO HANG PLANTS FROM THE CEILING WITHOUT DRILLING HOLES

After making your plant hanger, you will be dying to decorate your home by hanging your plants from the ceiling. The problem is that many plants are quite big and heavy.

Here are some good ways to hang as many plants as you want to create a warm and welcoming environment and save space, without drilling holes into the ceiling.

Magnetic plant hooks

Magnetic plant hanger hooks are great for any type of surface without ruining it. They come in stainless steel, nickel, chrome, or bronze finish and they are ideal for any type of decor.

As the name suggests, they work thanks to so-called rare-earth magnets (Neodymium-Ferro-Boron) and represent the most technologically advanced solution on the market today. Indeed, they can support weights up to 100 pounds.

For the magnet to work to its full potential, it should be attached directly to a clean metal surface.

Granted, finding a metal anchor point on the ceiling may not be easy. You should be looking for metal plates, bolts, large screws, or braces.

For instance, plasterboard screws are often found in homes, usually painted over to hide them. Simply glide the magnet over the ceiling to locate them.

But be careful, because the smaller the surface area of the anchor point, the smaller the magnet's ability to support the weight.

Steel wire rope or tension rods

Steel wire ropes running from one wall to another give the curious optical illusion of the plants hanging from the ceiling.

If the walls are close enough, you can also use a tension rod like the ones used for shower curtains. They are adjustable rods with springs inside and generally encased in a round plastic or metal bar on the outside.

They are durable, easy to use, do not require installation, and won't cause any damage to the wall.

Standard shower tension rods can reach from 41 to 72 inches. Their load capacity varies, but typically it ranges from 10 to 30 pounds, depending on the rod.

Extra long tension rods meant for room divider curtains can be 120 or even 160 inches – long enough to cover the width of a small room.

Just remember that as a rule, the rod must be at least 2 inches longer than the distance between the walls.

For plant hangers, shower rods are preferable because – being sturdier – they have a greater load-bearing capacity than curtain rods.

Self-adhesive hooks

These are hooks that stick to the wall thanks to their adhesive foam backing. Typically, they are used to hang keys or kitchen utensils; but you can also utilize them to hang plants from the ceiling without the need for nails or screws.

They work best if installed on a smooth clean surface. They can hold about a pound in weight. They are very convenient because you can stick them virtually anywhere you want.

The only caveat is that since they are meant to hold objects resting against the wall, you need to make sure that the plant hanger is securely attached to the hook.

Suction cup hooks

Suction cup hooks are another good alternative to installing your plant hangers without drilling holes in the ceiling.

The only flaw is that they only work on perfectly smooth surfaces. Therefore, walls are not ideal candidates, but skylights or other glass, metal, or plastic surfaces will do the trick.

Their load capacity is limited – typically about one pound – although some models on the market boast capacities up to 10 pounds. The best suction cup hooks feature a swivel hook that can rotate to fit both flat or sloping ceilings.

Bar clamps

Bar clamps are tools used by carpenters. They consist of an adjustable jaw that slides on a metal guide. Thanks to this mechanism, they can be secured onto wall ends (up to 4 inches thick) to support your plant hangers.

C-clamps

If there are exposed beams in your room, then C-clamps may be your best option. In fact, they can be tightened firmly onto the beam to support a considerable weight – up to 50 pounds for a 6-8 inch clamp.

To prevent damage to the wooden beam, buy rubber pads to apply to the clamp jaws.

The best plants to hang

The above are the best solutions for hanging heavy plants from the ceiling without drilling or damaging your walls. However, if you do not have reliable anchor points, your only option is to keep the weight of the hanging plant to a minimum.

The most lightweight ones are those that do not need soil. Here is a list of stylish plants that have the virtue of being extremely lightweight:

- Air plants (Tillandsias)
- Anthurium andraeanum
- Basil
- Lucky bamboo (Sanderiana Dracaena)
- Monstera deliciosa
- Peace lily
- Philodendron wend imbe
- Pothos
- Spanish moss
- Umbrella plant.

Chapter 3

OUTDOOR PLANT HANGER

Textile waterproofing method

Plant hangers made from natural fibers are not very suitable for outdoor use, because they gradually spoil when exposed to the elements.

The ideal materials for outdoor plant hangers are synthetic fibers such as nylon or polypropylene.

If you still want to use natural fiber cords for your outdoor plant hangers, there is a method you can use to waterproof the material.

What you need to use is wax.

Here is how to proceed:

- Melt 4 ounces of paraffin wax and 4 ounces of beeswax pellets using a double boiler.
- Using a small brush, coat your project with the melted wax, one section at a time.
- Dry the waxed part with a hair dryer – this will bond the wax to the fabric.
- Alternatively, place the waxed plant hanger inside an old pillowcase and tumble dry for about 15 minutes.

Cleaning your plant hanger

To clean your outdoor plant hanger, all you need to do is:

- remove the plant
- clean with mild detergent and cold water
- allow drying thoroughly.

Chapter 4

3 IDEAS TO JAZZ UP YOUR PLANT HANGER

1. Dyeing your plant hanger

Would you like to add some color to your plant hanger? Try dip-dyeing. Here's how to do it in 5 easy steps:

1. Purchase some textile dye and proceed with the preparation following the instructions on the box (they will surely be there).

Use a glass jar or wide bowl to prepare the color, so it doesn't spill.

2. Slowly dip your plant hanger starting with the bottom and until you reach about one-third of its length.
3. Now, hang the plant hanger onto a hook or dowel so that it hangs above the glass bowl or jar.

This is to allow the soaked part to absorb the dye well.

Allow soaking for about 30 minutes.

4. Detach the plant hanger and rinse in warm water to remove any excess dye.

Rinse until the water comes out clear.

5. Hang to dry.

2. Add colorful thread inserts

Adding colorful embroidery thread is another creative idea to decorate your plant hanger. Here's how:

1. Grab embroidery thread pieces of different colors
2. Wrap some sections of your plant hanger with the thread to your desired length.
3. When you're done, hide the thread ends inside the wrapped parts using a toothpick, needle, or stick.
4. Do the same with different colors.

3. Add mini terrariums

An original alternative to the classic potted plants is mini terrariums.

They are great little hanging gardens that you can create and customize.

Day after day, you will be able to observe the growing sprouts as they soak up all the light filtering through the glass bowl.

What you need:

- a glass container (can also be an old jar)
- pozzolana (volcanic rock) for water drainage
- small gravel
- decorative sand
- a mix of soil and sand
- decorative elements (such as stones, pebbles, moss, and lichens)
- plants of your choice (indoor plants only) such as Phytonia, Echeveria, or Chlorophytum.

How to make a mini terrarium:

1. sterilize the container by washing it with hot water and allow it to dry well
2. place the first layer that will serve as drainage
3. lay the soil
4. dig holes in the soil to insert the plant sprouts
5. add decorations
6. water.

Maintenance:

Your mini terrarium requires little maintenance. If you seal it with a cork stopper, a water cycle will form spontaneously, and it will only need to be watered 3 times a year with a small amount of water.

You only need to make sure that the substrate stays moist.

The terrarium needs to be in a well-lit spot, yet not exposed to direct sunlight.

The ideal temperature is between 15°C and 27°C. If the temperature is higher, simply uncork.

The same applies if you notice that excess condensation.

For drainage, add a thin layer of about 3 cm (1.20 inch) of pozzolana to the bottom of the pot.

Chapter 5

4 BEST MACRAMÉ PLANT HANGERS

Project #1 DIY Spiral Macramé Plant Hanger

Supplies

- 3mm (0.12 inch) macramé cords of different lengths:
- 6 pcs x 140" (inner cords)
- 2 pcs x 170" (outer cords)
- 2 pcs x 24" (wrapping knots)
- 1 wooden ring with a diameter of 2"
- sharp scissors
- 5-inch plant pot (between 4 and 6 inches is ok).

Method

STEP 1: Thread the 8 cords through the wooden ring. The two longer cords should be one on the left and one on the right, with all the others in between.

STEP 2: Tie the cords with spiral knots holding 2 long cords on the left as one cord and the two longer cords on the right as the other cord.

Position the left cords in front of the central ones to create a shape similar to a number 4.

STEP 3: Guide the right cords under the left ones and then behind the center cords. Then pull the ends through the loop you find on the left.

Pull both cords to tighten up the knot.

STEP 4: Make a series of 25 spiral knots (if you don't remember how to tie a spiral knot, you can find it in the first book).

STEP 5: Take one of the shorter (24") cord pieces and start tying a wrapping knot. Hold the cord against the others to form a U shape, with the short end sticking up on the left.

STEP 6: Wrap the long end of the cord 6 times around the other cords and complete the wrapping knot by pulling the ends of the cord.

Step 7: Cut the 2 pieces of excess cord from the wrapping knot.

STEP 8: Measure the cords about 12" down from the wrapping knot.

STEP 9: At a distance of 12", tie 4 square knots to create the top of the plant hanger area.

STEP 10: Measure 3 more inches down and tie 2 square knots between the previous 4.

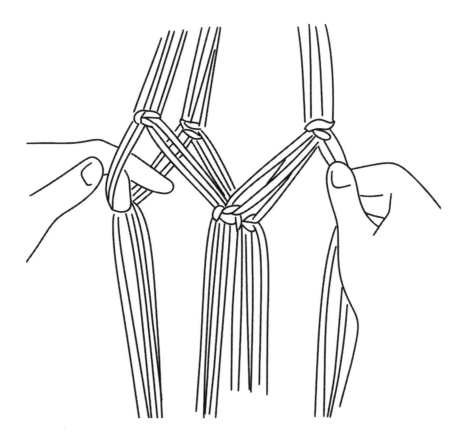

STEP 11: Repeat all around by making 4 alternating square knots.

STEP 12: Before tying the final wrapping knot, make sure that the plant pot sits well in place.

STEP 13: Take the last piece of 24" cord and finish with another wrapping knot.

STEP 14: Cut the cords about 5-6" from below the bottom wrapping knot. If you like, untwist the cords and comb them to form a fringe.

Project #2 DIY Diamond Plant Hanger

Learn all the steps required to create a beautiful macramé with the diamond pattern technique (aka diamond knots).

Supplies:

- 3 mm (0.12") 3-ply twisted macramé cord:
- 10 pcs x 134" cords
- 1 pc x 50" cord (for the wrapping knot)
- 8" wooden dowel
- plant pot 5" wide by 5.5" tall
- sharp scissors

STEP 1: Grab the longer cords, fold them in half and attach them to the dowel with lark's head knots. Attach them by spacing them out evenly.

For convenience, you can hang the dowel with two S hooks to a coat rack.

STEP 2: Start creating the left side of the diamond pattern.

Take the two center cords. Cross the left cord over the right to form a 4-like shape.

Wrap the end of the left cord around and thread it through the resulting loop.

Take the next cord to the left and repeat the process again, tying a half hitch knot around the right cord.

Hold the two cords together, take the next cord to the left and tie a half hitch knot around the two cords using the leftmost cord.

Now hold the three cords together, take the next cord to the left and tie another half hitch knot around the three cords.

STEP 3: Repeat the pattern until you reach the last cord on the left.

STEP 4: Grab the last cord and bring it behind the bundle of knotted cords.

STEP 5: Make one last half hitch knot, but this one should be reversed.

STEP 6: To complete the left half of the diamond, tie a half hitch knot with each cord, going back towards the center.

Step 7: Now, proceed with the right side of the diamond.

Take the two center-right cords.

Wrap the rightmost cord around the leftmost cord in the shape of a 4. Thread the end into the loop.

Holding the 2 cords together, grab the next one to the right and do the same thing.

Repeat until you reach the last cord on the right.

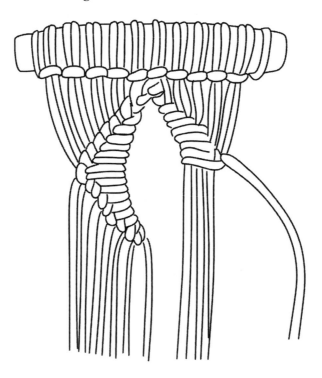

STEP 8: As before, wrap the cords with the last one on the right.

Repeat the knots until you return to the center.

STEP 9: Connect everything to form the diamond pattern.

Take the right center cord and wrap it around the left-center one. Tie a half hitch knot to connect them together.

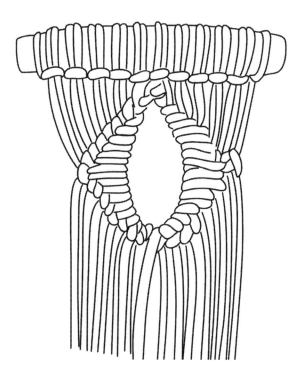

STEP 10: Repeat the whole diamond pattern 2 more times.

STEP 11: Proceed to tie the square knots for the base of the plant hanger.

STEP 12: Measure a distance of about 5" from the last diamond.

Make a row of 4 square knots, skipping the 4 outermost cords (2 on the right and 2 on the left).

STEP 13: Grab these 4 cords and join them in the middle by knotting them into a square knot about 1" below.

STEP 14: Use the 2 cords left from the last square knot and the two cords to the left of the first group to tie another square knot about 1.5" below the knot you just tied in the center. It'll be on the left side.

STEP 15: Repeat on the right side.

STEP 16: Keep making square knots all around at the same level.

This is what it should look like.

STEP 17: Take the 50" piece of cord and tie a wrapping knot around all the cords tied with the square knots.

STEP 18: Cut the unwanted excess cords and finish by creating a fringe.

Project #3 DIY Hanging Fruit Basket

This macramé project that you are going to make is multi-purpose. You can use it both as a plant hanger and as a hanging fruit basket.

Supplies

- 3mm (0.12") 3-ply twisted cotton cords:
- 8 pcs x approx. 126" (320 cm) for the main working cords
- 1 pc x 33" (approx. 84 cm) for the top wrapping knot
- 16 pcs x approx. 67" (170 cm) for the basket
- 1 pc x 50" (127 cm) for the bottom wrapping knot
- 2 pcs x 8" wooden rings
- sharp scissors

Method

STEP 1: Take the 8 working cords and fold them in half.

STEP 2: For the upper part of the basket, make a braid and fold it into a loop secured with a wrapping knot.

STEP 3: Tie a temporary knot and then start braiding.

STEP 4: To make the braid, separate the cords into 4 sections, two by two.

Move the second segment over the first one, from right to left. Then shift the fourth section over the third, again from right to left. At the end, the second over the third.

STEP 5: Continue tying the four-strand braid until it reaches the desired length (approx. 6-7").

STEP 6: Untie the overhand knot that held the braid together (be careful not to undo the braid), fold the braid, and tie a wrapping knot with the 33" cord. Your basket hook is done.

Step 7: Now, grab 4 strands and tie a square knot with them about 1" down from the wrapping knot.

STEP 8: Bring the working cords behind the center cords and move the center cords to the outside.

STEP 9: Make another square knot about 3" down. This is called a switch knot.

STEP 10: Tie 4 more switch knots, about 3" apart going down.

STEP 11: Repeat for the other 3 arms of the basket.

STEP 12: Now attach the first wooden ring. Do so by tying DHHKs around the ring.

First attach the second cord, then the first one, then the fourth one, then the third one.

STEP 13: Repeat the process for the other 3 arms, spacing them evenly around the circle.

STEP 14: Now proceed to attach the 67" cords.

Attach them to the wooden ring by tying a lark's head knot between the basket arms.

Take the right end of the cord you just attached and knot it onto the wooden ring.

Do the same thing with the left cord on the other side.

Repeat with the remaining cords.

STEP 15: Proceed to create the body of the basket.

Make 3 rows of alternating square knots. Start about 1" down from the ring.

After you have done the first row, do the second, at a distance of another inch.

Finish with the third row, further down by another inch.

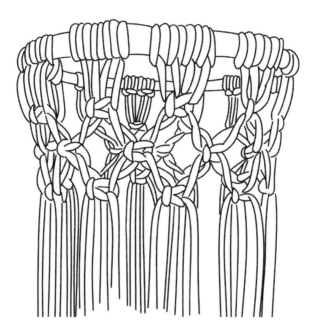

STEP 16: Proceed to attach the second wooden ring.

Simply attach the cords in the same order as they are with a double half hitch knot.

STEP 17: To close the bottom of the basket, you need to collect all the cords together.

With the last remaining cord, tie a wrapping knot around the cord bundle.

STEP 18: Cut any unwanted length of the cords and make a fringe if you want.

Project #4 DIY Double Plant Hanger

This project allows you to have two sturdy plant hangers in one! It's great if you are a plant lover but have little space in the house.

Here are all the steps you need to follow to complete it.

Supplies

- 3 mm (0.12") twisted cotton cord, cut as follows:
- 6 pcs x 188" (approx. 478 cm) for the inner cords
- 2 pcs x 212" (approx. 538 cm) for the outer cords
- 3 pcs x 24" (approx. 61 cm) for the wrapping knots
- sharp scissors
- measuring tape

If you want, you can use cords of different colors to provide vibrancy and add wooden beads between some of the knots.

Method

STEP 1: Take the 8 longest pieces of cord and line them up by leveling one end.

Put the six 188" pieces in the center and the two longest (212") cords on either side.

STEP 2: Fold the cords in half and find the center of the shorter ones. Hold the center point with one hand and place the cords on a work surface.

STEP 3: Grab the tape measure and measure 3" from the center point. Tape down the cords where you measured the 3" mark.

STEP 4: Tie 15 square knots starting from the tape. Use the two outer cords as working cords.

STEP 5: After completing the knot sequence, un-tape the cords. Fold the knots sequence you just made to form a loop.

STEP 6: Take one of the 24" cords and tie a wrapping knot.

Step 7: Once you have finished, hang the loop with an S hook on your clothes rack (please refer back to the section dedicated to the perfect macramé workstation).

STEP 8: Make the arms of the upper section of the plant hanger.

Separate the cords into 4 sections of 4 cords each.

STEP 9: Take the first section and tie 22 spiral half knots (if you don't remember how to make them, you can find the directions in the first book).

STEP 10: Repeat the 22 half spiral knots with all the other 3 sections.

STEP 11: Now grab the tape measure and measure 5" down from the end of the spiral knots on one of the arms.

STEP 12: With the cords of that section, tie 8 square knots at a distance of 5" from the end of the spiral.

STEP 13: Repeat the last 3 steps with the other 3 sections.

STEP 14: Take the tape measure again, and measure another 5" down on one arm.

STEP 15: Now join two adjacent arms.

Take the 2 right cords from the left-hand arm and the 2 left cords from the right-hand arm and connect them with a square knot at the 5" mark.

STEP 16: Connect the other arms of the plant hanger in the same way.

STEP 17: Take the tape measure and measure 3" from the last row of square knots.

Make another row of square knots alternating with this one.

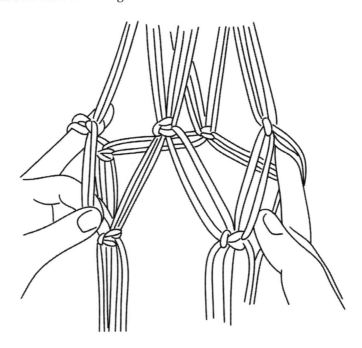

STEP 18: Leave another 3" gap and close the cords with a wrapping knot using another 24" cord.

STEP 19: To build the second part of the plant hanger, all you need to do is just repeat all the steps.

So:

• divide the cords into 4 sections of 4 cords each

- tie 4 sets of 22 half spiral knots (distancing them as before)
- tie 4 sets of 8 square knots (replicating the previous spacing)
- make two alternating rows of square knots.

STEP 20: Tie everything together with a wrapping knot using the last 24" piece of cord.

For a finishing touch, you can simply trim the excess cords into a tassel or unwrap them into a fringe.

Conclusions

Plant hangers are among the best-loved boho chic macramé items.

Now that you have learned how to make them with your own hands, you are ready to jump head first into the most fashionable and trendiest aspect of macramé.

In the fourth book, we are going to explore the boho chic trend that is enjoying huge success all over the world and see how to decorate your home based on the principles of this new vogue. Plus, you will have the opportunity to experiment with many new projects.

Don't miss all this (and much more) in the next book!

BOOK 4
MODERN MACRAMÉ: BOHO CHIC

**Remember to check out your Easy Download
and Print Instructions with Extra Pictures
for all the Projects in this Book**

Please scan the QR code or follow the link on Page 166

Introduction
MACRAMÉ MANIA: A TREND-SETTING ART FORM

In the first three books, you got to the heart of macramé, starting from the basics and continuing to learning advanced techniques. This allowed you to create objects that you never thought could come from your own hands. The gratification was instant, wasn't it?

Now it is time to explore one of the most interesting aspects of this art: how it has modernized so much to become a trend, indeed a web 'catchphrase'.

In fact, macramé is now one of the most popular 'arts and crafts' on social networking sites, such as Instagram and YouTube. One of the reasons for its success is that it lends itself perfectly to the *Do it Yourself* phenomenon for which the public literally goes crazy.

Another reason is that it is an extremely creative textile art with which everything can be made, from objects for home and garden decor to jewelry, bags, accessories, even clothing.

The charm of macramé lies precisely in this: a millenary heritage that has led this extremely versatile technique to adapt to all the eras and peoples with whom it has come into contact.

In recent years, we are witnessing a great revival of macramé - one of the pivotal elements of the *boho chic* style - in textile workshops all over the world.

Contemporary macramé is characterized by plenty new patterns that skillfully combine ancient techniques with modern taste.

So much that it was able to conquer even the biggest international haute couture catwalks, becoming a true evergreen.

Chapter 1

MACRAMÉ AND BOHO CHIC: THE ORIGINS

From hippie craftsmanship to haute couture

The boho chic style has close ties to the hippie fashion of the 1970s and is characterized by the use of natural fabrics and vintage motifs, combined with trendy accessories.

Because of their common characteristics, macramé has integrated seamlessly into boho chic, becoming almost a hallmark of this style.

But boho is not just a current. It is a real way of life that stems from an ideology that in the 1960s and 1970s could be described as 'rupture'.

Today, however, it has fully entered the mainstream. The wide variety of boho clothing and accessories have made it a phenomenon that has given rise to an artistic, 'relaxed' and free fashion, with loose, light clothing and casual accessories.

Thanks to macramé, boho chic has exploded not only in clothing, but also in home and room decor in general.

Contemporary macramé is in fact adaptable to many types of décor, from country to industrial, shabby-chic to Nordic style.

What does boho mean?

Boho is a contraction of *Bohème, a* term for a non-conformist lifestyle.

The Bohemian style is known as distinctive type of fashion, very different from the usual trends. In fact, this definition applies to the entire history of the trend.

More than 200 years ago, bohemian was a term referring to an exotic style, usually associated with the artists of the time, as well as eccentric writers and intellectuals. But in the beginning, the true representatives of this style, the *bohemians*, were those who belonged to nomadic or even refugee peoples.

Soon, however, the term was associated to the most disparate personalities who did not fit into any social class.

Everything that has to do with bohemianism is associated with freedom and non-conformism, without forgetting adherence to deep-rooted and important values. This is why influential personalities throughout history were also associated with the category. We speak, for example, of Sherlock Holmes, William Shakespeare and Winston Churchill.

The name *Bohemian* is also associated with the cultural period that arose after the French Revolution and brought more libertarian and egalitarian thinking. The era saw many artists plummet into poverty due to the deprivation of the previous system of patronage that guaranteed them a secure income.

The Bohemian Current

The first mention of the Bohemians dates back to the end of the 18th century, during the period of the French Revolution. At that time, due to the hostile social and economic climate, artists and creative people were destined to a life of poverty. Reduced almost to hardship, they were forced to make do and wear second-hand and old clothes.

This circumstance left artists, even when things improved later on, with the desire to express their creative side through clothing and eccentric ways.

The name *Bohemian* derives from the fact that ordinary people associated the artists' way of dressing as similar to that of nomadic gypsies, who were in fact originally from a European region called Bohemia.

Hence the term *bohemian* (later contracted into boho) became synonymous with the counterculture associated with creativity and contempt for traditional, conformist values.

The hippie era

The Bohemian movement came to life again in the 1960s when it was the hippies who opposed a conventional lifestyle.

The resulting new style of clothing - including ethnic dresses, embroidery, macramé, mixed prints, volumes, fringes - began to become popular and also changed fashion forever.

The hippies, in their rejection of traditional values, were also radical in their fashion choices that made them favor loose, 'relaxed' clothes instead of the sleek, slim silhouettes that were all the rage in the previous decade.

Boho Chic Macramé: art that respects the environment

Macramé is an art that puts respect for the environment first. The concept of sustainability in fashion was also introduced thanks to the example of this ancient technique.

Macramé is the exact opposite of disposable fashion, which is cheap but spoils quickly and above all is made from polluting materials. Macramé works are aesthetically valuable and in addition they last because they are made from first-class materials.

The use of organic and natural fibers, characteristic of macramé, is less harmful to the environment because it does not require the production of chemical or toxic substances.

Many suppliers, among others, are keen to bring to market organic cotton ropes produced from natural seeds and without the use of pesticides and other harmful chemicals, but with the controlled use of pest-killing insects.

Cotton produced this way is softer, more workable and resistant, and consequently lasts longer.

Furthermore, it is biodegradable. Any cotton fiber that cannot be recycled or used further can be composted and therefore does not take up space in landfills. In a compost heap, cotton biodegrades in about 5 months, although in some cases a week may even be enough (especially if it is 100% cotton).

MACRAMÉ AT HOME: THE CHARM OF BOHO CHIC

5 ideas for a boho chic home with macramé

From tapestries to beautiful plant hangers and delightful table runners, here are 5 tips to give your home a boho chic touch with macramé.

1.   Create a warm atmosphere with tapestries on the walls

Made of woven cotton ropes usually starting from a wooden support, such as a branch or stick, they give any room an exquisitely bohemian touch.

Perfect in the living room but also in the bedroom, to enhance the walls. Depending on the desired effect, you can hang just one, as long as it is the right size for the space you intend to decorate, or more than one. There are tapestries of various shapes, developed vertically or horizontally, both monochrome and in different colors.

2.   Create energizing green corners with plant hangers

They can be singles, doubles or bunk beds, plant hangers - as you have seen in book 3 - give a touch of vitality and energy that is indispensable for any home.

Plant hangers not only allow you to use the space vertically - which is perfect especially if you have small rooms - they also create a scenic effect given by the green of the plants enhanced by the neutral color of the plant hanger itself.

You can combine different plants; the important thing is to concentrate them in certain places in the house so as not to create clutter.

You can hang plant hangers in the living room, the kitchen but also in the bedroom.

3.   Give your table an original touch with tablecloths or runners

The kitchen is another corner of the house that you can enhance with macramé.

Tablecloths and runners are just some of the items you can use.

Runners, for example, lend themselves as the basis for elegant centerpieces that you can make yourself.

The only caveat is not to overdo the colors, keeping them neutral is the ideal choice.

Equally elegant and with vintage charm are macramé coasters or coasters that you can also use as candle holders or for other decorative elements.

4.   Make the living room special with cushions, chair decorations and hammocks

Chairs and stools decorated in macramé are the elements that most of all give your living room a boho chic touch.

You can also decorate poufs or cushions and alternate them on the sofa, perhaps with a blanket or sofa cover, again in macramé.

Equally delightful are hammocks. Although they are mainly used outdoors, you can make models that are also suitable for indoors. Just choose the right size for your room and you will have a truly unique corner for relaxing and reading.

Installing a hammock in the house is not difficult, you just need to use supports that are appropriate for the weight. The hammock also allows you to utilize space vertically, as it is suspended from the ceiling.

5.   Use the charm of lampshades to make the living room more chic

The macramé lampshade, besides being very romantic, is the ideal decorative element to cover an old lamp and give it a unique vintage charm.

You can also use macramé lampshades to customize your bedside lamps. When you turn on the lamps, enjoy the intertwining threads that create a spectacular play of light and give the room a magical touch in perfect boho chic style.

The 10 best color combinations for a boho-style home

Would you like a super-cool home that exudes a cozy, relaxed energy?

The answer is interior design in pure boho style.

Houses that follow this style are immediately recognizable, because they have a globally inspired décor that often revisits traditional design rules to create an original and personalized look.

In the previous section, we looked at how macramé décor objects can be placed in rooms. Now we will deal with the colors.

What are the boho colors?

There are actually no fixed rules when it comes to boho. If the set of colors you have chosen recalls the element of air, earth, open spaces, then you are following the boho style.

The basic colors are brown, blue, red and green, although many prefer neutral shades.

These colors are often combined with deep purple and orange.

All mixed well with pieces of furniture made with the macramé technique.

Imagination and creativity have free rein here, there are no mistakes, but only indulging our desire for freedom.

Below are 10 examples of combinations in pure boho chic style.

1.   Neutral and earthy colors

Neutral colors combined with earthy tones are a classic combination. Just think of ivory, light grey and cream combined with warm browns and slate grey.

Light-colored walls make paintings, plant hangers and tapestries stand out better. They also reflect natural light very well and provide an excellent basis for unleashing creativity.

2.   Terracotta and green colors

Rural houses are beautiful and convey serenity and peace because they use colors close to nature.

Terracotta is a fantastic earth tone that gives a sense of calm, warmth and serenity. In addition, it is a color that goes beautifully with different shades of green.

3.   Green and warm brown colors

The color green stands out in most boho-style homes and is often found combined with warm brown tones.

Green corners can be achieved with plant hangers. You can create real living walls.

The color green gives a fresh touch and goes perfectly with vintage furniture, armchairs with floral upholstery, tapestries and carpets in warm colors.

4.   Blue and ivory colors

In this case, the dominant inspiration is ocean hues combined with a neutral color such as ivory that gives a fresh, almost Greek island feel to your home.

5.   Mustard yellow and orange colors

Yellow, which has always been associated with sunshine and vitality, is not an extreme choice if introduced wisely into the home. However, if you feel particularly daring and are in the mood for something truly original, you can experiment with the combination of mustard yellow and a lighter shade of orange.

Try it for example in the bedroom. Give it a finishing touch with a macramé bedspread in earthy tones and geometric patterns.

6. Deep red and off-white colors

Oxblood red and burgundy are among the most brilliant and intense shades when combined with off-white.

You could, for instance, think of the living room wall in red with macramé curtains in creamy white.

7. Neutral and wooden colors

You can add neutral colors to your walls with tapestries or other macramé accessories.

Combine this with wooden floors and you will create a wonderful color combination!

8. Grey and pink colors

This combination of dark grey and cheerful pink is especially ideal in children's rooms.

The ambience can be made even more joyful with vintage toys and brightly colored carpets.

9. Colors mint green and black

An unusual color combination but one that can give great satisfaction. Provided it is balanced with functional accessories and furniture.

10. Blue and gold colors

Classic blue combined with gold accents gives a *boho-luxe* look. Different shades of blue in the room, combined with brass accessories are one of the trendiest boho color combinations.

Chapter 3

MACRAMÉ JEWELRY FOR A BOHO CHIC LOOK

The 6 best types of rope for making macramé jewelry

In the 1980s, macramé jewelry was very fashionable. Hemp in shades of earth, beige, brown and rust was the standard.

Waxed cotton also existed, but only in these same few shades.

Then synthetic braided cords were introduced, but they were too thick for jewelry. So, if you wanted some color, the solution was embroidery thread, which was pretty, but not very durable.

Today, in the midst of a new trend, the options for jewelry strings are many. Here is a small guide to choosing the best ones.

1. Hemp ropes for jewelry

New production techniques have introduced softer hemp cords onto the market and they are available in a range of bright colors, although hemp in its natural color retains its charm.

Hemp cable is available on cards or reels in thicknesses of 0.5/1/1.5mm (0.02/0.04/0.06 inch).

The new hemp cords are excellent for macramé jewelry.

Be careful when choosing, because thicknesses may be indicated by weight. For example, the 10-pound cable is about 1 mm, the 20-pound cable is about 1.5 mm, and so on.

You can find the cable in raw hemp for a more rustic and natural look, or coated in wax.

2. Bamboo ropes for jewelry

Bamboo has recently arrived as a sustainable alternative to synthetic fibers. In fact, this material is famous for being solid and very productive.

Its fibers are so silky and strong that it is perfect for use in textiles and beyond.

They are available in craft shops in a wide variety of colors and thicknesses.

Bamboo cord is often coated with wax to make it smooth and shiny. while it effectively holds the knots pretty well, it maintains sufficient flexibility for bead crochet and the famous kumihimo. The thickness most commonly used for jewelry is 1 mm or 1.5 mm (0.04/0.06 inch).

3. Braided cotton cord for jewelry

Cotton cord is similar to bamboo and hemp cord. It is often sold in craft shops wrapped on cards of different sizes.

Especially suitable for jewelry, it can be natural or wax-finished.

It is available in a wide range of colors and its thickness varies from 0.5mm to 2mm (0.02/0.08 inch).

4. Linen lanyard for jewelry

Linen is obtained from the *Linum usitatissimum* plant. Although it may look similar to hemp or cotton, it is very different from these materials and is an excellent alternative for making macramé jewelry.

The size of the linen cord is indicated by the number of 'folds', i.e. the number of threads grouped together.

The most commonly used dimensions for jewelry making are 3 layers (approx. 0.55 mm/0.02 inch), 4 layers (approx. 0.7 mm/0.03 inch) and 7 layers (approx. 1.2 mm/0.05 inch).

5. Chinese jewelry cord

The Chinese knot string is made of synthetic material: a braided nylon that is silky to the touch and holds knots well.

Available in a wide range of thicknesses and colors, e.g. 0.8 mm (0.03 inch), 1 mm (0.04 inch) and 1.5 mm (0.06 inch).

The only downside is that the finish of the Chinese knot string can be slippery.

6. Nylon C-Lon and S-Lon cords

The C-Lon is a 3-ply braided nylon cord approximately 0.5 mm (0.02 inch) thick. An excellent size for making micro-macramé jewelry.

C-Lon is available in more than 100 colors, from bright to dim to fluorescent.

C-Lon Tex 400 is a slightly thicker wire: approximately 0.9 mm (0.04 inch) in diameter. Although thicker, it is also very suitable for making macramé jewelry. In fact, the cable is very silky and easy to knot.

S-Lon is a very similar product to C-Lon. It is available in more or less the same colors and thicknesses, so it can be easily interchanged.

The step-by-step technique for placing a stone in your jewelry

Macramé jewelry in pure boho chic style? Insert a stone, crystal or *cabochon*.

Beautiful to embellish your jewelry, they can also be used as decorative elements to enrich wall hangings or other furnishings.

Here we will show you the quickest and easiest technique to firmly insert stones or crystals into your macramé.

Materials

* A stone, crystal or cabochon of your choice
* cotton ropes (max. 2mm thickness/ 0.08 inch):
* 4 strings approx. 60 inch (152.4 cm) long
* 1 rope approx. 20 inch (50.8 cm) long
* craft glue
* sharp scissors.

Procedure

STEP 1: Take the first rope and bend it to form a loop.

STEP 2: Take the stone of your choice. Squeeze the ring at one end to take the measurements. The ring must be large enough to fit the stone well. Hold the point with your fingernail and tie a knot to close the ring to the appropriate size.

STEP 3: Take the other 3 ropes and tie them to the ring with lark's head knots.

Make sure the distance is equivalent.

Test the stone to make sure it fits perfectly.

STEP 4: Once you have wrapped the end of the stone with the ropes, stretch them over the stone and decide at what height you want to tie the other row of knots.

STEP 5: Hold 2 strings close together. Remove the stone from the strings and tie the first knot at finger height. This first knot will act as a guide for the other 3.

When you have finished the knots, check one more time that the stone is firmly attached to the net you have made.

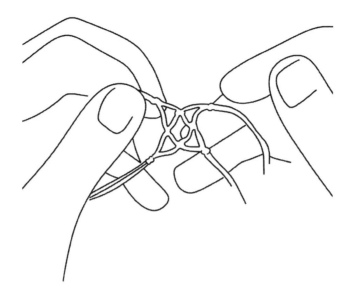

STEP 6: Repeat the last step for the other rows of nodes. The number of rows will depend on the size of the stone you have chosen.

STEP 7: When you have reached the point of covering the stone completely with the net, close it with a final knot.

You can tie a single sailor's knot with all the ropes.

Remember that the knot must be firm and tight.

DIY Macramé tassel earrings

Materials

- 4 cotton ropes 4 mm (0.16 inch) thick by 12 inches (approx. 30 cm) long
- 2 jute cords, 12 inches (approx. 30 cm) long
- 2 jump rings
- 2 hooks for earrings
- jewelry pliers.

Procedure

First of all, to help you out, use a macramé board or a cork panel as a support.

STEP 1: Attach the jump ring to the macramé board or cork board.

STEP 2: Thread 2 cotton ropes through the jump ring so that they are perpendicular to each other. The ring should be in the center.

STEP 3: tie a crown knot (see book 2).

Take the lower cable at the top right and place it over the right cable.

Take the right cable and bring it over the top cable with the end pointing to the left.

Take the upper cable and place it on the left.

Take the left cord, bend it to the right and pass the end through the loop formed at the beginning.

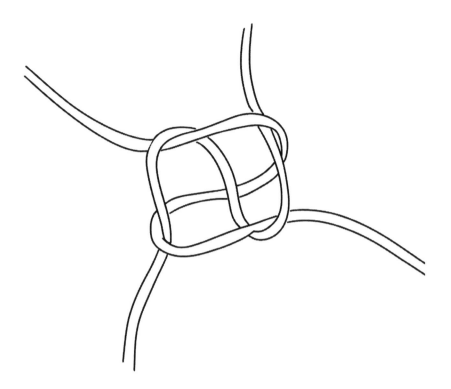

STEP 4: Pull the ends of the cable together to form a crown knot.

STEP 5: Use the same procedure to tie 4 more crown knots.

STEP 6: Take a piece of jute twine and tie a wrapping knot underneath to close the crown knots. Cut off excess jute.

STEP 7: Use the jewelry pliers to open the jump ring and insert an earring hook into it.

STEP 8: Cut the tassel and if you wish, brush it to create a fringe.

STEP 9: Repeat the same process to make the other earring.

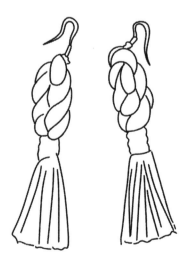

$$\mathcal{C}\!hapter\ 3$$

THE BEST BOHO CHIC FURNITURE

DIY Boho Macramé Curtain Wall Hanging

Materials

- White cotton macramé rope (any color you like) braided in 3-ply 3 mm (0.12 inches) of the following length:
- 70 cables of 20 feet (approx. 6 m)
- 36-inch support rod (approx. 92 m)
- Scissors.

STEP 1: Divide the ropes in half and attach them all to the pole with a lark's head knot.

To support the rod and easily tie knots, use a clothes hanger.

Together the strings should cover at least 30 inches of the rod.

STEP 2: Make 5 triangles using square knots.

Each triangle will have a diameter of 7 square knots on the first row to 1 on the seventh row.

Use alternating square knots as you see in the photo.

STEP 3: Add half nodes.

This addition to the pattern gives a dynamic touch to the design.

Take four ropes and tie them 2 by 2 together with a half knot.

This is a curtain that is simple to make and at the same time very elegant. You can use it in so many ways.

In order for the curtain to open easily allowing you to pass through, you must not tie the knots all the way.

Going downwards, they must thin out.

STEP 4: Cut the ropes to the length you want.

DIY Boho Macramé Mirror Wall Hanging

Materials

- 4 mm (0.16 inch) macramé rope in the following size:
- 4 strings of 108 inch (approx. 2.8 m)
- octagonal mirror
- 1 2-inch (approx. 5 cm) wooden ring
- 4 x 25 mm (approx. 1 inch) wooden beads with 10 mm (0.4 inch) diameter holes
- sharp scissors.

Procedure

STEP 1: Fold the 4 wooden ropes in half and knot them all to the wooden ring with a lark's head knot.

Make sure you tie the knots tightly.

STEP 2: Separate 2 of the ropes from the others. Make a first square knot with these.

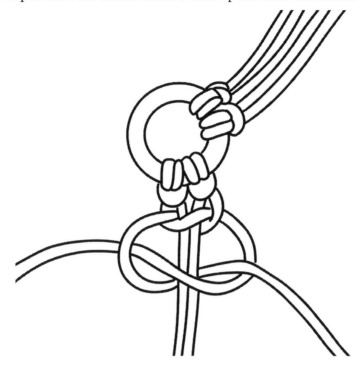

STEP 3: tie another square knot on the other two ropes as well.

STEP 4: Tie the two square knots together as you see in the picture below.

STEP 5: Make a series of 7 square knots with all 4 strings by connecting them as you have seen in the picture.

STEP 6: Divide the remaining free strings like this: 2 strings on the left, 4 strings in the middle, 2 on the right.

Tape the ends of the cables together.

This will make it easy to string the beads.

STEP 7: Thread a bead onto the two left strings. Do the same with the right strings.

Close with a knot under the bead.

STEP 8: Tie the 4 strings in the center with a simple wrap-around knot (about 0.5 cm/0.20 inches) under the beads.

STEP 9: Take a rope from the center-left and tie it to the 2 ropes on the left.

Do the same to the right.

STEP 10: Add the mirror and check that the height of the nodes is uniform.

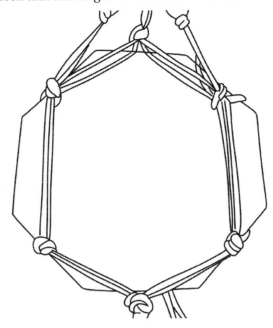

STEP 11: Add one of the 3 side cords to the back of the mirror to hold it firmly in place.

STEP 12: Tie all 3 side strings at the bottom left and right of the mirror with simple knots.

Separate the 3 side strings again. Send 1 of the strings from each side to the back of the mirror while the other 2 remain in front.

Tie all the ropes. Tighten the front knot.

STEP 13: Turn the mirror upside down and tie the ropes at the back together.

Tie all the ropes tightly. Tighten the back knot.

STEP 14: Fray the ends of the ropes.

DIY Boho Macramé Placemats/Table Runner

With the technique you are about to learn, you can also make wonderful table runners by simply varying the length of the strings.

Materials

* 20 single-strand 3mm (0.12 inch) cotton ropes measuring 114 inches (approx. 290 cm)
* 1 wooden dowel (which will be removed later. You only need it to hang the ropes and work better)
* scissors.

Procedure

STEP 1: Attach all the ropes to the wooden dowel and secure them with a lark's head knot (you can hang the dowel on a coat hanger).

STEP 2: Make 3 rows of alternating square knots.

Start with the first line a few centimeters from the dowel.

Make sure you leave some space between the lines as you see in the picture.

STEP 3: Start the diamond (or rhombus) pattern.

Start from the left. Jump 2 ropes and tie a square knot.

Skip four more ropes and tie a square knot.

Repeat until the line is finished.

Leave the last 2 strings out as well.

STEP 4: Make a normal row of square knots along the length of the row.

STEP 5: If you have followed all the steps in step 6, you should find large empty spaces. Make 4 square knots to reconnect those spaces.

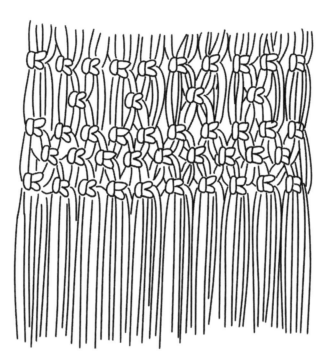

STEP 6: For the next line, repeat what you did in STEP 3.

STEP 7: Repeat the pattern (STEP 3 and STEP 5) until you have a total of 22 lines.

STEP 8: Close the pattern with 3 more rows of alternating square knots (as you did at the beginning).

STEP 9: Remove the placemat from the dowel.

STEP 10: Open the loops of the lark's head knots by cutting them with scissors.

STEP 11: Hang the placemat in half and cut the excess string to even out the fringe on both sides of the placemat.

CONCLUSIONS

Be inspired by the 10 most talented macramé artists

The best way to get better and better at macramé is to be inspired by the best. This will not only help you think of new ideas the time, but above all it will help you develop an above-average sense and taste for aesthetics.

What is the benchmark for macramé artists? We have already said it: it is all Instagram.

Think of the posts with #macrame numbering over 3 million. Not considering all the other hashtags such as: #macrametutorial, #macramejewelry, #macramemakers, #modernmacrame.

On Instagram you find real macramé stars followed by hundreds of thousands of people. For an art that can be described as niche, these are respectable numbers.

If you are wondering which are the best ones to follow, here is a list of the 10 most talented macramé artists. You can admire their work and be inspired by following their profiles on Instagram:

1. The Dutch Rianne Aarts has created a small empire with her Teddy and Wool. Her e-commerce is really rich in proposals for those who love home decor and natural fibers.

She was asked to create decorative macramé furniture for Google offices, Hilton Hotels and Starbucks coffee shops.

Her style is characterized by the creation of large, striking decorative wall panels made by playing with elaborate patterns and nuances perfectly designed for any type of environment.

2. JAC is Jamie Cother's macramé brand. She mainly makes home decorations that she also teaches to others in her many workshops.
3. Bobbiny is the Polish artist who sells her macramé creations worldwide. Her works range from home decorations to more 'functional' objects such as pet beds or bookmarks.
4. Agnes Hansella, a Jakarta-based artist, is famous for creating monumental works but also delightful pieces only a few centimetres in size.
5. Dutch artist Sandra de Groot creates real soft sculptures in the form of elaborate wearable headdresses and decorated armor-like planes.
6. Katie from Stella Blue Boho uses natural and locally sourced materials to make her macramé pieces. She buys string and crystals from local shops and driftwood from local beaches.
7. Windy Chien claims to dedicate itself to the exploration of a macramé knot every year. Noteworthy are her modern-style chandeliers and tapestries.
8. Michael Gabrielle is famous for making pieces measuring 25x30 feet. He is also an illustrator and sculptor.
9. Shian Ellen started her small home craft business after facing a health problem that prevented her from working. She says that macramé has helped her through very difficult times.
10. Elsie Goodwin is famous for the geometric balance of the patterns in her tapestries. She also emphasizes the therapeutic aspect of macramé in addition to the creative aspect. She worked in fashion before turning her hand to macramé.

BOOK 5
CHRISTMAS MACRAMÉ

**Remember to check out your Easy Download
and Print Instructions with Extra Pictures
for all the Projects in this Book**

Please scan the QR code or follow the link on Page 166

INTRODUCTION

Rediscovering the true spirit of Christmas with macramé

Every year in December all over the world there is a tradition that brings along something magical: the custom of Christmas decorations. Trees, balls, garlands, everything contributes to creating a warm and cozy atmosphere.

Every home at Christmas, even the simplest and most modest, is filled with colors, scents and lights. There is a healthy enjoyment in the anticipation and preparation for Christmas. Even adults become children again and are able to get away from all the worries of everyday life for a few hours. Perhaps because it is one of the few times when there is a sense of peace and harmony. Or, at least, that's how it should be.

At this time of year, macramé is a big player. Not only because of its ability to make the atmosphere even more enchanting. But because it brings us back to the true spirit of Christmas. If you only think about what this art represents, a great symbol of unity through knots, you realize that it fully reflects the meaning of Christmas: being united in harmony, creating a wonderful work in which every single knot is important.

Thanks to macramé, you don't have to spend a lot of money, but will have just simple things that you have created yourself with your own hands, perhaps helped by your children.

You can also choose to recycle objects and decorate them with macramé to turn them into decorations you like.

Between the search for that special gift and the frenzy to organize the perfect Christmas for the family, we often lose sight of the opportunity to take a step back, take time for ourselves, and reflect - especially during the festive season - on the impact of our actions on the environment.

Macramé gives you the opportunity to:

* stop and rediscover the true spirit of Christmas
* creating environmental-friendly Christmas decorations
* give original, eco-friendly gifts to family and friends.

This book is full of tips for experiencing Christmas more consciously thanks to macramé. You will discover how to decorate your home in a unique way and there will be plenty of gift ideas for those you love.

Chapter 1
4+2 SURPRISING CHRISTMAS DECORATIONS

DIY Christmas balls

Materials

- colored Christmas balls or wooden ball
- 8 mm ecru cord:
- 10 cords of approx. 60 cm (23.62 inch)
- 1 lanyard approx. 50 cm (19.68 inch)
- scissors
- pencil.

Procedure

STEP 1: Cut 10 approx. 60 cm (23.62 inch) strings of macramé cord.

STEP 2: Gather 9 ropes and fold them in half. Take the tenth string and wrap it to create a circle.

STEP 3: Braid the 9 strings around the circle you formed with the tenth with a lark's head knot.

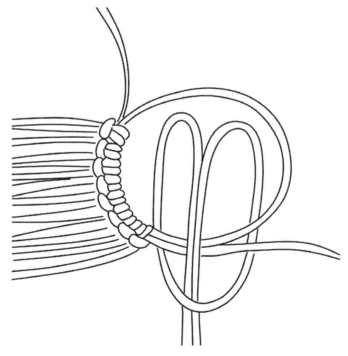

STEP 4: Pull the two ends of the ring cable and tighten it around the ball.

STEP 5: tie a square knot around two ropes and continue all around tying a total of 5 square knots.

STEP 6: Make the square knots leaving a space between each knot. You can use a stick or a knitting needle so that the space you leave is even for all 5 square knots.

STEP 7: tie another row of square knots around the ball, alternating with the previous ones to create a diamond pattern. Always use a stick to space the knots evenly.

STEP 8: Repeat the step twice more (always using the same spacer). You should have 3 groups of square knots as a result.

STEP 9: Use the 50 cm (19.68 inch) rope to wrap the ends of the other ropes with a wrapping knot.

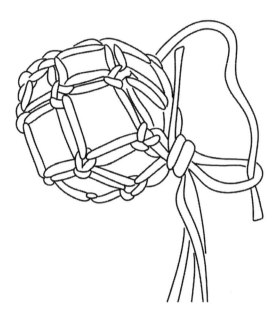

STEP 10: Cut all strings at a 8 cm (3.15 inch) distance.

DIY Christmas star

The star is the main symbol of Christmas. It cannot be missing as a decorative element in so many variations.

Here is a quick and easy method to make beautiful stars to hang in your home.

Materials

- a wooden or metal ring as a central base
- 3 mm (0.12 inch) macramé rope in a length of:
- 21 cotton ropes of approximately 1 meter (3.28 ft - the length of the ropes always depends on the size of the project)
- wooden beads
- scissors
- glue.

Procedure

STEP 1: Knot 20 threads around the wooden hoop with a lark's head knot so that they cover the entire circumference of the hoop.

STEP 2: Divide the strings into 5 groups of 4.

STEP 3: Continue the pattern with square knots.

STEP 4: Close the tips with a half hitch knot.

STEP 5: string wooden beads on the tips

STEP 6: Finish by cutting off the excess threads. You can decide to leave small fringes for a snowflake effect or fold them over the back of the pattern and fix them with glue.

STEP 7: Take the remaining cable and tie it to the top of the loop with a lark's head knot.

STEP 8: string two more beads divided by knots to finish the cable, which you will need to hang the finished star.

DIY Christmas tree

Materials

* Twisted cotton macramé rope in a length of:
* 1 14 inch rope (approx. 35 cm)
* 14 10 inch strings (approx. 25 cm)
* wooden beads
* comb
* scissors.

Procedure

STEP 1: Fold the 14-inch rope in half and insert a bead through the loop you have thus created. Leave a space of about 1 inch between the loop and the bead.

STEP 2: Take the two 10-inch strings and create a flat knot.

The flat knot is very simple, below you will find the diagram for its realization.

STEP 3: Take another 12 strings and tie them together at the center string with the flat knot you have just learnt.

Remember that if you want to increase the size of the tree, just use more rope, and tie more rows of knots.

STEP 4: Take the scissors and cut the strings to the left and right to give a triangular shape.

You can leave the central string a little longer and close it with another bead if you wish.

STEP 5: Use the comb to unravel the cables and form fringes.

Adjust the threads again to give the tree its final shape.

Add hairspray or starch spray to further fix the fringe.

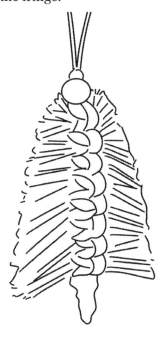

DIY Christmas wreath and more

With wooden circles of all sizes covered in macramé, you can first create simple and quick decorations to hang on the Christmas tree or, as a group, on a wooden dowel, to hang on the wall.

But there's more. The process we show you here is the same one that allows you to create - with a larger wooden or metal ring - a beautiful **Christmas wreath**.

Again, depending on the diameter of the ring and the thickness of the wire, you can also make:

* funny **napkin rings**
* beautiful **earrings** or **necklace decorations**
* **handbag rings** and... whatever your imagination suggests.

Materials

* wooden circles
* cotton cord (3 mm - 0.12 inch is fine).

Procedure

STEP 1: Take the wooden ring and cotton rope and start tying the rope to the ring with a lark's head knot.

STEP 2: Take the left end of the thread, loop it and tighten to form a knot.

Repeat until you have covered half the ring.

STEP 3: Take the right end of the thread, loop it and tighten to form a knot.

Repeat until you have covered the other half of the ring.

STEP 4: Close the ends of the ropes with a simple knot and cut off the excess ropes.

STEP 5: Take another thread that you will need in case you want to hang the wooden ring thus covered and knot it with a lark's head knot.

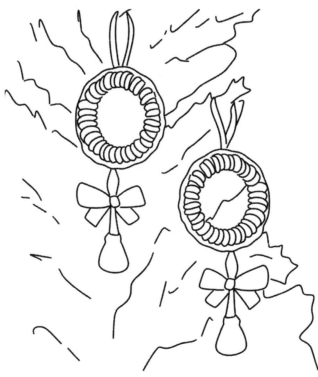

BONUS 1: DIY macramé owl tutorial

The macramé owl was one of the iconic macramé items in the 1970s. Here we offer it in a version that fits perfectly with the Christmas atmosphere.

Materials

* 3-strand cotton cord 3 mm (0.12 inch) to be cut according to these measurements:
* 8 strings 94 inch (2.40 m)
* 1 x 106 inch (2.70 m) rope
* 22 5 inch (12.7 cm) strings
* 2 x 20 inch (50.8 cm) strings
* 1 wooden dowel of at least 5 inches (12.7 cm)
* 2 metal rings (key rings)
* 2 wooden beads (with the hole to fit the thickness of the string)
* 1 short stick for the perch (at least 3 inches - approx. 8 cm)
* sharpened scissors
* comb
* hairspray or starch spray (optional).

Procedure

STEP 1: Fold the 106-inch rope in half.

Slide your fingers so that both sides are at 5 inches (about 12 cm) distance from the center.

Drop the rope on the left side and hold it with your right hand only. The result is a rope with the left side 5 inches longer.

STEP 2: Attach the rope on your right hand to the dowel with a lark's head knot.

STEP 3: Bring the left rope to the left side of the dowel, over the dowel. Pass it over the dowel and thread the end through the loop in the middle.

Bring the end of the cable behind the dowel and over, while holding the cable to form a loop. Pull the end through the loop and secure it tightly.

Between one knot and the other there should be about 8 inches (about 20 cm).

STEP 4: Attach the eight 94-inch ropes to the main cable with lark's head knots.

STEP 5: tie a square knot with the four center strings.

Make a square knot to the left and one to the right of the central square knot.

Make a row with two more square knots alternating with the first three.

Close with a square knot in the next row, in the center.

STEP 6: Take the last cable on the left and make 8 DHHKs going down to the center.

STEP 7: Do the same with the right-hand cable.

STEP 8: Again, take the left outer cable and make 8 DHHKs running down from the left to the center.

STEP 9: Do the same with the right external cable.

STEP 10: Take the seventh string starting from the left and do 2 DHHKs going down the center.

STEP 11: Take the seventh rope from the right and do 3 DHHKs going down from the right towards the center.

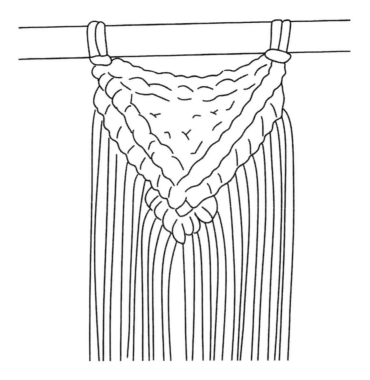

STEP 12: string the wooden beads you got on the fifth string from the left and right.

STEP 13: Make a square knot with strings 6-7-8-9 (starting from the left). Make another square knot with the next strings on the right.

Close the face with another square knot on the next row immediately below.

STEP 14: Take the two outer ropes on the left. Make 10 vertical lark's head knots using the outer rope as the working rope and the other as the filling rope.

STEP 15: Make the same string on the right side by reversing the function of the strings.

STEP 16: Take the center cable on the left, use it as a filler cable and make 8 DHHK going to the left. The last knots must meet with the row of vertical lark's head knots you made on the left.

STEP 17: Repeat the same thing, but to the right, so take the central cable to the right and make another 8 DHHKs that go to the right and meet up with the row of vertical lark's head knots you made earlier.

STEP 18: Take the 4 center strings and use them to make 2 square knots in a row (one on top of the other).

Make a square knot to the left and a square knot to the right of the two central knots (as seen in the photo below).

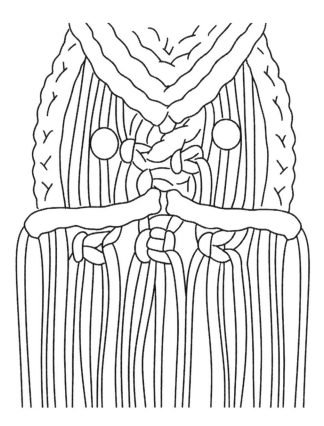

STEP 19: Under the row of 3 square knots make a row of 4 square knots alternating with the previous one.

STEP 20: Make another row of 3 square knots alternating with the previous ones.

STEP 21: Make a final row to close 2 square knots.

STEP 22: Take the outer cable to the left and make 5 DHHKs going down from the left towards the center. Leave an empty space before doing the 5 DHHKs (see next photo).

Do the same thing on the right.

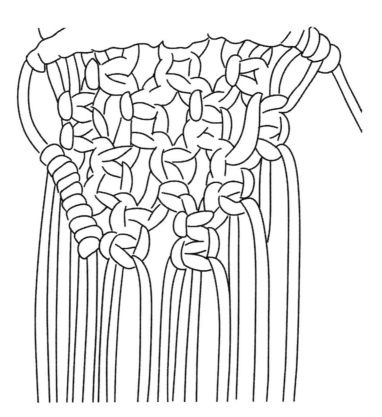

STEP 23: Take the 4 center strings on the left (from the square knot on the left) and tie another 5 square knots. Repeat the same step on the right (see photo).

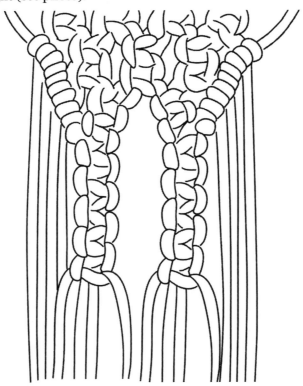

STEP 24: *This step-by-step procedure is used to create the owl's eyes.*

* Take the two metal rings that you have previously prepared.
* Take one of the 20-inch (50.8 cm) pieces of string. Fold it so that the right side is 3.5 inch (8.9 cm), leaving the left side longer.

- Attach the rope to the metal ring with a lark's head knot at the point on the rope that you have identified.
- Tie 5 vertical lark's head knots around the ring going to the left.
- Take a 5-inch (12.7 cm) rope, fold it in half and fasten it to the metal ring with a lark's head knot. Repeat the operation 5 more times.
- Do the same procedure to create the other eye.
- Leave the two long ropes intact because you will need them to attach the owl's eyes.
- Unravel and fray all the other ropes.

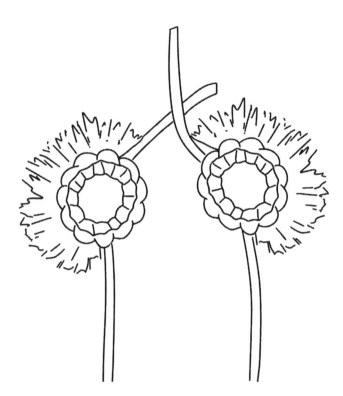

STEP 25: *This step-by-step procedure is for attaching the owl's eyes.*

- Place the eyes you have just made with the metal rings at the level of the beads you have previously placed in the pattern so that the beads are in the center of the rings.
- Work on the back of the pattern.
- Bring the dangling eye strings to the back and tie the 2 strings with a double knot (see photo).
- Do the same with the other eye.

STEP 26: Do you remember that before you did the last row of vertical lark's head knots on both the right and left side, you left some space? You will need this space to create the wings of the owl.

Take 10 5-inch (12.7 cm) cables. Attach to the rope you left uncovered 5 cables on the left with lark's head knots and 5 cables on the right.

STEP 27: Now that you have made the owl's snout and body, all that remains is to attach the little legs to the perch.

Take the wooden stick.

Take the 2 ropes to the right under the row of square knots. Bring them in front of the wooden stick that serves as your perch and wrap the row of square knots around it.

Fasten a square knot beneath the perch to give moer stability.

Repeat the same operation with the row of square knots on the left.

STEP 28: You can finally proceed with the finishing touches.

Draw the strings you used to make the flaps (both left and right) together with the scissors.

Fringe with a comb. If you consider it necessary, spray a little hairspray or starch spray to fix the fringe more securely.

STEP 29: Here's the finishing touch! You should have one of the longer ropes left over. Use it to gather the dangling strings with a wrapping knot.

Cut the end cords thus gathered to the length you want, and unroll them to form a tassel.

Your beautiful owl is complete: congratulations!

BONUS 2: DIY macramé stocking

Materials

* 5 mm cotton rope
* 2 pieces of fabric (cotton, jute, whatever suits you best for a Christmas stocking)
* 1 macramè board or stand
* pins
* glue
* sewing machine.

This is the only project among those we have presented that requires a sewing machine. This is because it would have been too long and complicated to make a stocking entirely in macramé. Therefore, we have come up with the idea of making a stocking with a fabric base to which the macramé pattern will be applied on the front.

If you do not have a sewing machine, you can use a needle and thread.

As far as the pattern that will decorate the stocking is concerned, here we give you some indications, but if you feel particularly creative, with the notions you have learnt so far you are perfectly capable of creating a pattern of your own.

STEP 1: Place the two pieces of fabric on the backing and secure them with drawing pins (or needles).

Cut out pieces of fabric to give the shape of a stocking.

Join the two pieces of fabric with needle and thread or with a sewing machine.

Turn the sock inside out so that the seams are on the inside and cannot be seen.

You have prepared the base of your stocking.

STEP 2: Cut 10 ropes 8 times the height of the stocking.

STEP 3: Loosen the seam at the top of the stocking to pass the rope folded in half one by one. Fasten it to the thread with a lark's head knot.

Repeat with the other 9 strings.

STEP 4: Once you have a row of 8 strings attached with lark's head knots, start making a row of 5 square knots.

STEP 5: Make a successive row of square knots, alternating with the previous one.

STEP 6: Repeat the pattern for two more rows.

STEP 7: Close the upper pattern of the sock with a row of HHKs to create a line effect.

STEP 8: Continue the central part of the pattern with a diamond pattern (as you saw in the first book).

You can invent as many patterns as you like to decorate your stocking.

STEP 9: To close the pattern, you can opt for a simple row of HHKs going down vertically with the leftmost string.

Once you get to the curved part of the foot, add more rope by simply creating a small hole at the top of the sock big enough to insert the rope.

STEP 10: Make another row of HHKs descending vertically with the rightmost rope.

Once you get to the bottom, do another row of HHKs to close the foot of the sock.

STEP 11: Cut off the excess rope. Using hot glue, glue the remaining pieces of rope to the foot of the sock.

STEP 12: Add a thread to the top of the stocking to hang it.

Chapter 2

3 SPECIAL ITEMS FOR A PERFECT CHRISTMAS TABLE

DIY macramé candle holder

Materials

- Cotton macramé rope (should be about 4 times the height of the jar):
- 47 24-inch strings (approx. 60 cm)
- 1 x 34-inch rope (approx. 86 cm)
- glass jar 6-inch (approx. 15 cm) high
- scissors.

Procedure

STEP 1: Take the longest rope and attach all the others folded in half with a lark's head knot.

STEP 2: Once all the ropes are knotted, attach the longer rope to the edge of the jar with a simple knot. The ends of the longer rope will be used as a final pair of ropes.

STEP 3: Choose 2 pairs of ropes to serve as centers and start with a square knot.

Continue with 2 more square knots.

Repeat around the entire circumference of the jar.

STEP 4: Add a round of square knots between each row.

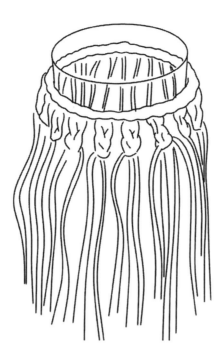

STEP 5: Complete the pattern tip with HHKs for each string group.

STEP 6: Adjust the length of the ropes with the scissors and tension them so that they are stretched perfectly.

DIY quick macramé coasters

This design is suitable to be used as a coaster, under cup, under plate or even as a saucer. The speed lies in the thickness of the rope.

Materials

- 8 mm (approx. 0.3 inch) thick macramé rope of this length:
- 1 6-foot rope (approx. 183 cm)
- 5 x 48-inch strings (approx. 122 cm)
- 6 x 36-inch strings (approx. 91 cm)
- 2 x 12-inch strings (approx. 30 cm)
- macramé board or any other support preferably cork
- T-needles
- sharpened scissors
- crochet needle
- fringe brush
- cup as a reference model for measurements.

STEP 1: Take the 6-foot rope, place it on the support and make a circle with one end.

STEP 2: Attach the five 48-inch ropes to the loop you created with an inverted lark's head knot (see book 2).

STEP 3: Tighten the ends of the long rope to close the loop.

STEP 4: Take a T-shaped needle and attach the work to the cork board.

STEP 5: Proceed with 13 double half knots to the left in a spiral. Use the long rope as filling cord.

STEP 6: Take one of the 36-inch ropes, fold it in half and attach it to the filling cord (the longer rope) with an inverted lark's head knot.

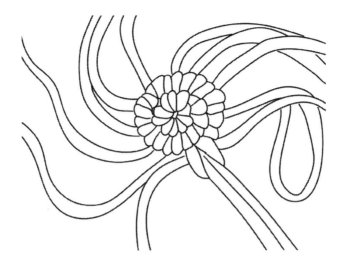

STEP 7: Continue with double half knots until another large space is created between the strings. Then add another rope by attaching it with an inverted lark's head knot.

Continue tying double half knots and adding extra strings (finish the longest ones first) until the coaster is the size you want (use the cup as the measurement).

STEP 8: Turn the coaster upside down and cut the excess thread coming out of the center.

STEP 9: Take the crochet needle and thread the ends of the strings through.

Thread the ends under some knots in the back of the coaster.

Cut the excess rope.

STEP 10: Last step, create the fringe with the brush.

DIY Christmas table runner

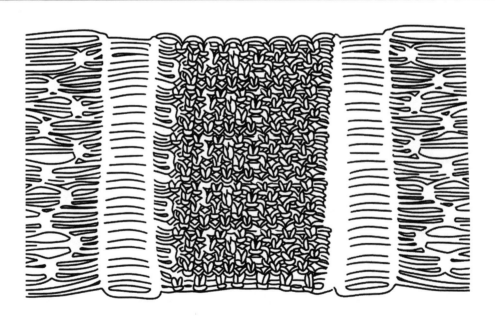

Materials

- 12 inch (approx. 30 cm) wooden dowel
- 3 mm thick cotton rope (0.12 inch) cut into:
- 22 strings of 16 inches each (approx. 56 cm)

- hooks for hanging the dowel
- scissors.

Procedure

STEP 1: Fold the ropes in half and hang them on the dowel with a lark's head knot. You should have 44 ropes to work with.

STEP 2: Descend 6 inches (approx. 15 cm) and use the outermost string to the right as a filler string to make a series of horizontal HHKs along all 44 strings, forming a line from right to left.

STEP 3: Start again from the right, leaving a space of approximately 1.5 inch (3.8 cm). Use the 4 rightmost strings to make a square knot.

Leave the four strings on the left free, and then tie another square knot.

Repeat until the line is completed.

STEP 4: Make a subsequent row of square knots alternating with the previous one about 3 inches (7.6 cm) apart.

STEP 5: Leave an additional 11-inch (approx. 28 cm) gap and leave the two outer strings on the right and two on the left free.

STEP 6: Loosen another 4 ropes from the right and tie a row of square knots.

STEP 7: Make another row of square knots alternating with the previous one 3 inches (7.6 cm) apart.

STEP 8: Make another row of HHKs starting from the right to the left.

Go down approximately 2.5 inches (6.35 cm) and use the same basic rope to create another horizontal row of half knots from right to left.

STEP 9: Go down about 1 inch (2.5 cm) and start this time from the left with a row of square knots.

Leave no spaces and continue with another line of square knots alternating with the first.

Repeat until you have 13 rows of alternating square knots without spaces.

STEP 10: Repeat step 8.

STEP 11: Repeat from step 5 to step 8.

STEP 12: Cut the ends of the rope at the level you wish but so that they are of equal length.

STEP 13: Remove the strings from the dowel by gently pulling out all knots. Cut the center of the lark's head knot.

Adjust the outer strings so that they are all the same length.

3 ORIGINAL GIFT IDEAS FOR CHRISTMAS

You can turn all the projects you have learnt so far into beautiful gifts. Here we present 3 more ideas for special Christmas gifts for family and friends.

DIY macramé key chain bracelet

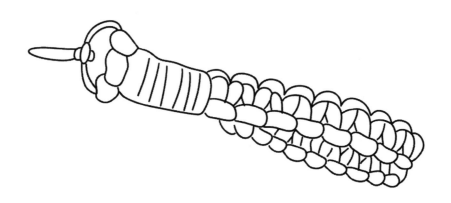

Materials

- 5 mm (approx. 0.2 inch) cotton macramé cord of this length:
- 2 x 65-inch strings (approx. 165 cm)
- 1 x 30-inch rope (approx. 76 cm)
- metal key ring carabiner
- glue
- scissors
- T-pins
- macramé board or stand.

Procedure

STEP 1: Take the carabiner to close the key ring and attach the two 65-inch ropes with a lark's head knot.

Do not fold them in half. The two outer ropes must be much longer.

Take the first rope and measure 14 inches (approx. 35.5 cm). Attach the rope to the carabiner at that point.

Repeat the operation with the other string.

Make sure the shortest parts are in the middle.

STEP 2: Fasten the carabiner and ropes to the macramé board (or cork board).

From the clasp of the karabiner, go down about 2 cm (about 0.8 inch) and start tying a square knot.

STEP 3: Continue making a row of square knots until only a few centimeters of rope remain.

Do a trial run to make sure the length of the knots fits your wrist (or that of the person you want to give it to).

STEP 4: Add fabric glue in the space between the karabiner and the row of knots.

Create a circle with the row of knots and glue the end where you applied the fabric glue.

STEP 5: Finish by taking the smallest cable and making a wrap-around knot as in the photo.

STEP 6: Finish by cutting the ends of the rope that stick out.

DIY macramé wine bag

Materials

- 4 mm (0.16 inch) 3-strand twisted cotton rope in the following lengths:
- 2 x 40-inch strings (approx. 102 cm)
- 12 strings of 66 inch (approx. 168 cm)
- 1 x 30 inch (76 cm) rope
- scissors
- cork/macramé stand
- T-needles
- hooks and hangers to hang your project.

Procedure

STEP 1: Take one 40-inch rope and fold it in half. Link the two ends of the rope with a simple knot.

Take the other 40-inch rope and do the same.

STEP 2: Overlap the two ropes thus folded to create a loop in the center as in the photo.

STEP 3: Knot the 12 66-ich ropes with lark's head knots to the ropes thus folded, 6 ropes above and 6 below.

STEP 4: Hang the project on a coat rack with hooks.

STEP 5: Starting from the top, leave about 1 inch (about 2.5 cm) of space and tie a row of square knots, 3 on one side and 3 on the other.

STEP 6: Leave another inch of space and make another row of square knots, alternating with the previous one.

Continue the pattern until you have done 8 rows (or until you have reached the height you desire).

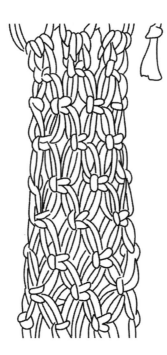

STEP 7: tie a wrapping knot of at least 5 turns of thread just below the last row of square knots.

STEP 8: Cut the excess threads to the length you want to form a tassel.

DIY paracord pet collar in macramé

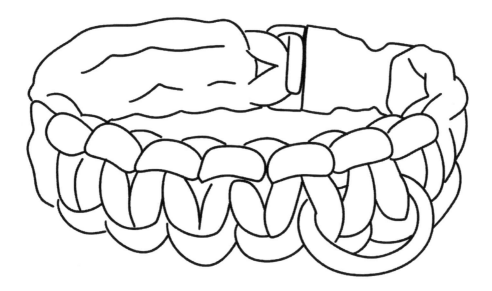

Materials

*   paracord of the following length: 1 ft of cord per inch. For example, if you want your collar to be 15 inches long, you will need 14 ft of paracord
*   a 0.5-inch buckle (plastic or metal). For small dogs you can use a 0.5-inch buckle. Half size 0.75 inch up to 3 inch. The size of the buckle also influences the design of the pattern. For example, on the 0.5-inch buckle there is only room for two strings
*   a ¾ inch D-ring.

Procedure

STEP 1: Cut the rope to the length you want for your collar (see instructions above) and fold it in half.

STEP 2: Flatten the ends of the rope with a lighter and pliers, so that they fit over the buckle.

STEP 3: Thread the end of the rope through a buckle attachment and tie a lark's head knot through the buckle.

STEP 4: Thread the other ends of the string through the other buckle attachment. Leave enough space according to the size of your pet's neck.

STEP 5: From underneath the buckle attachment, start making a series of SKs until you fill half the space you left.

STEP 6: Take the D-ring (which will be used to attach the carabiner for the leash).

Thread the ring through the outer rope on the left and insert it in 3 SK so that it is securely attached to the collar.

STEP 7: Continue with the series of SKs until you have completed the entire size of the collar.

STEP 8: If you can, check the measurements by having the dog wear it before cutting the 2 remaining outer strings.

Cut off the ends of the cable and melt the cable with a lighter, so that it does not fray.

CONCLUSIONS

Are you looking forward to December to put everything you have learnt in this fifth book into practice? For you, Christmas will never be the same again!

Now you are able to make beautiful Christmas decorations on a low budget, plus they are environmentally friendly. And that's not all, the projects you found inside the book are perfect for making as a family with twice the fun!

This journey, which has taken you to explore the infinite universe of macramé, is coming to an end. But it is not over yet.

Speaking of making something with your family, we have come up with one last unmissable book for you: macramé for children.

Continue reading to discover all the benefits of macramé for little ones and many wonderful projects for them to do.

BOOK 6
MACRAMÉ FOR CHILDREN

**Remember to check out your Easy Download
and Print Instructions with Extra Pictures
for all the Projects in this Book**

Please scan the QR code or follow the link on Page 166

INTRODUCTION

At what age should you start macramé?

Did you know that Maria Montessori - the famous Italian pedagogist made famous around the world for the educational method that bears her name - was already proposing looms for children aged 2-3 years to practice tying knots on?

They are very special frames that still exist today and can be purchased online.

The well-known scholar believed that the activity of knotting was a real workout for the child's mind and body. Ties zips and buttons were for her interactive games that presented him with challenges essential to his growth.

Today, it is rare for a child starting primary school to be able to tie his own shoes. Due to sneakers with Velcro and little patience, knots have been almost completely eliminated from childhood. It is not uncommon for a kid to reach the age of 8-9 without being independent.

This, however, risks impairing abilities related to so-called *fine motor skills*, i.e., motor control over small movements of the hands and fingers and the small muscles of the face, mouth, and feet.

If the child has difficulty with this type of movement, this can lead to frustration and a lowering of his or her self-esteem because he or she cannot do 'things'.

It is true that smartphones and video games are also suitable for the stimulation of 'fine motor skills, but their misuse could even generate serious side effects, such as eye problems and postural damage, up to real disorders in the social sphere (addiction, reality distortion, anxiety, and depression).

That is why an activity such as macramé is extremely important for a child's healthy growth and development. Macramé not only teaches children to make complex movements with their little fingers, but it also stimulates their imagination and creativity. And all without contraindications.

And since macramé does not require any tools that can harm them, they can also start practicing this art from the age of 4-5, with the help of an adult.

Chapter 1

ALL THE BENEFITS OF MACRAMÉ FOR CHILDREN

7 reasons to learn macramé from an early age

All children need to be stimulated from an early age to develop their full potential. And everyone is creative.

We often start by having them do small crafts with paper, glue, and scissors. As they grow, however, they need to discover other craft materials that they can transform with their unlimited imagination!

They discover themselves and the world around them by exploring, playing, and experimenting.

Macramé is not only a lot of fun; it is also an ideal activity for their development. It positively influences both cognitive (thinking, listening, learning), physical (as we have already seen), and emotional skills. It is also great for cultivating the very important skill of problem-solving, which boosts their self-confidence.

Our lives are so hectic and busy that we cannot find time to relax with our children. Making macramé projects with them is a perfect way to spend quality time together. Working closely together will create a greater connection with them and strengthen your bond.

Think also of the benefit of disconnecting, even if only for brief moments, from the constant workings of a mind constantly busy thinking about a thousand things.

Want to learn more? Here are 7 reasons to teach macramé to children and young people.

1. **Macramé stimulates *fine motor skills***

Fine motor skills are made up of small movements of the wrists, hands, fingers, mouth (including tongue), and feet. These movements require greater effort and concentration for the child.

By stimulating these skills, macramé enables the improvement of 'fine motor skills, which is achieved through a greater connection between the two hemispheres of the brain. This, consequently, also positively affects cognitive abilities such as learning, memory, and concentration.

Macramé is particularly suitable for fine motor skills because it trains children's fingers and eyesight, teaching them to juggle increasingly complex knots.

2. **Teaches how to recognize and match colors**

Children learn to recognize and name colors from one and a half to five years old.

Learning is facilitated by games and activities such as macramé. Just think of the infinite variety of string colors that can be combined with each other. In the project section, your children can learn - for example - how to create beautiful colored bracelets.

3. **Stimulates creativity and imagination**

Perhaps one of the most important benefits of macramé is that it provides ample room for the child's creativity. Through the art of knots, he will realize how many choices he has and how many resources he has at his disposal. He will also discover that there are no limits to his imagination and that he can create anything with a little good-will and patience.

4. **Teaches the value of collaboration**

Doing a macramé project together with an adult or other children and young people is good for the social and emotional development of children.

Learning to work together, share materials, and help each other greatly increases their empathy: they look at what the other is doing without envy or competition but with the pleasure of contributing to the greater good. This leads them to celebrate each other's successes with joy and learn the value of collaboration.

5. Boosts the child's self-esteem

Children take great pride in showing off their creations. That is why macramé helps them to strengthen their self-confidence.

Don't forget to praise your child when they have completed one of the projects in the book.

6. It is a safe and low-budget activity

Macramé is a very kid-friendly activity. No tools are needed other than strings and one's own hands. Of course, it may happen that you need scissors, but in that case, especially for young children, the steady presence of an adult is sufficient.

Another benefit is that it is a low-budget activity. All you need is some rope that you can get even from scrap, some simple patterns, and you're done!

7. It is a great alternative to smartphones and video games

Adults often struggle with being too chained in front of their PC, tablet, or smartphone. Social media is a wonderful way to keep in touch with our loved ones, but we must not overdo it.

Even the youngest children are not exempt from this problem. Video games and smartphones increasingly capture their attention, with a negative effect on their eyesight and physical fitness.

Better to put your mobile phone aside for a while and devote yourself to an activity that develops your child's motor, cognitive, and emotional skills without side effects.

The importance of learning by doing

Why is it that when we are children, we want to learn a thousand things while growing up it seems that learning new things is something we willingly avoid or only do if we must.

And even if we wish to embark on new paths, the first question we ask ourselves is:

"Why should I subject myself to the physical and emotional difficulty of new challenges?" "Why should I set myself bold, long-term goals if I have no guarantee of success?"

This limiting attitude in most adults has very deep roots and unfortunately stems from the fact that from childhood we were instilled with a learning method that favors theory over practice.

But this takes away all the fun, the sense of adventure, and discovery and leaves room for negative feelings such as fatigue, boredom, heaviness, and constraint. All things we try to avoid as we grow up.

That is why it is so important that from an early age that your child experiences the taste of learning by doing. Let him try, make mistakes, get angry and rejoice in his successes, and don't forget to always encourage him.

Hyperactive children? Try macramé

Did you know that art therapy is used as an alternative form of treatment in cases of severe hyperactivity such as ADHD (attention deficit hyperactivity disorder)?

The assumption behind the use of art (understood as craft) in cases of acute hyperactivity is that it helps to bring out emotions such as anxiety, sadness, or anger, which are sometimes difficult to express in words, especially for children.

Art therapy helps children who are more comfortable with images than words and prefer to communicate through artistic skills.

In particular, macramé work is an effective way to get your child to release excess energy. Following a pattern and keeping little hands busy seems to be particularly relaxing.

Here are some tips to get you started:

1. It is not the product that is important, but always maintaining a positive and motivating attitude.
2. Create a workspace with few distractions. Remove all electronic devices.
3. Have all the materials and workspace ready, so that your child can start without problems.
4. Start with something simple and the first few times, limit his choices (materials, patterns to follow) to prevent the indecision about what he wants to do from blocking him.

Chapter 2

ADVENTURE GUIDE WITH KNOTS

The 7 basic knots indispensable for camping, hiking, and outdoor games

What could be nicer for a child or teenager than to experience the outdoors, in close contact with nature, leaving TV, smartphones, and games aside for a few hours?

It has something magical, liberating, and relaxing about it. And it allows you to have real physical adventures, not virtual ones.

WARNING: This chapter contains instructions on how to tie some basic knots that are essential to know. This mini guide in no way replaces the guidance of an expert, which is indispensable for any beginner.

1. Half Hitch

The half knot is a simple sailor's knot. Insecure on its own, it is an excellent basis for other, much more reliable knots and hitches.

It is very simple to do. Just follow the diagram.

STEP 1: Wrap the rope around the support.

STEP 2: Pass the end through the loop and tighten.

USE: foundations and shoring, moorings, outdoor games, macramé projects.

2. Mooring Hitch Knot

Much safer than the Half Hitch Knot is a simple mooring hook consisting of one end that can be released easily.

Follow the diagram and instructions.

STEP 1: Wrap the rope around the pole and pass the end of the rope through the gap created.

STEP 2: Form a loop with the end of the rope.

STEP 3: Pass it through the loop behind the main rope.

STEP 4: Hold the strings together and pull to tighten.

USE: mooring a kayak or small boat for a short time, securing an animal, securing a tent, transporting or lifting something.

3. Overhand Knot

One of the simplest and most used knots. Below are the diagram and instructions.

STEP 1: Form a loop by passing one end of the rope under the other.

STEP 2: Thread the end of the rope into the loop and pull it out completely.

STEP 3: Pull the two ends of the rope tight.

USE: prevents fraying of the end of the rope or thread, excellent as a stopper and safety knot, used as a base for other knots, for outdoor games, and macramé projects.

4. Clove Hitch

Among knots with practical functions (nautical or hiking), it is considered one of the most important along with the *sheet bend* and *bowline*. It is always tied around a support.

STEP 1: Hang the rope on the support.

STEP 2: Loop the end of the string around the support.

STEP 3: Pass the garment behind the support.

STEP 4: Pull the rope tight to tighten the knot.

USE: to start or fix an anchor, to create a lock on a carabiner, to lift something, to tie a bandage in case of injury, in hammock suspension systems, to fix a tent, for outdoor games, and for macramé projects.

5. Bowline Knot

The bowline knot comes from the nautical world. Its use is very old and dates to the age of sailing boats in the 16th century.

STEP 1: Pass the end of the rope through the loop you created with the rope.

STEP 2: Pass the end around the other end of the rope and pass it back through the loop created.

STEP 3: Hold the two ends of the rope and pull the loop tight and tie the knot.

USE: for tying a mooring line around a mast, pole, anchor, or other. It is used in rescue operations. In rowing and sailing. In fishing, outdoor games such as climbing trees or tying kite strings.

6. Sheet Bend

Do you need more rope, but all you have is a thin thread? Solve that with a sheet bend knot. Also known as a 'weaver's knot', it is used to join two strings of different thicknesses but is also effective for strings of the same diameter.

STEP 1: Pass the thinner rope through the loop you previously made with the thicker rope.

STEP 2: Wrap the lanyard around the loop and under itself.

STEP 3: Hold the thicker end and the thinner rope and start pulling to tighten the knot.

7. Carrick Bend

The Carrick bend is a very solid knot suitable for joining 2 heavy ropes, elbows, or cables that are difficult to bend. Also called Josephine Knot (we talked about it in book 2), it is used to make original macramé motifs.

STEP 1: Take one of the two strings and make a simple loop with it.

STEP 2: Take the other rope and pass it through the 'arms' of the first rope (as seen in the picture).

STEP 3: Now loop the first string as if to form a weave.

STEP 4: Pull the ends of the ropes to close the knot.

USE: securely fixes heavy loads and is used in climbing. For hammocks, camping, and outdoor games.

6 fun knot games to play outdoors

Here are 6 fun games to teach children to tie their first knots. They are ideal to do outdoors and in groups. Fun is guaranteed!

1. Square knot relay

The square knot relay is a team game. Divide the group into two teams. Give each child a piece of rope.

Place a long piece of rope on the ground in the middle between the two teams.

At the signal, the first child runs to the rope lying on the ground, ties his piece of rope to the end of the other rope with a square knot (the explanation of the square knot can be found in the first book), then runs back to his team where he chooses the next child to do the same.

Each child in the team in turn must reach for the rope in the center and attach his or her own to that of his or her team with a square knot.

The first team to connect all the ropes with correct square knots wins.

2. The game 'Simone says

In this game, each child has a 3-foot-long piece of rope.

An external person (in this case an adult who also acts as a referee) pronounces the phrase 'Simone says' followed by the name of a knot.

Each child must perform the knot that has been named.

If there is no command 'Simone says ... (knot name)' the child does not have to do anything.

It is counted as a penalty if a knot is tied incorrectly, or if one knot is tied instead of another, or if it is tied at the wrong time.

After 3 penalties the player is eliminated from the game. Whoever is left last wins.

3. Blind knots

Here is another fun game! Again, you don't play in a group, but each on his or her own.

Make about ten different knots and put them in a cloth bag. Each child, blindfolded, will try to find as many knots as they can.

They can stick their hand inside the bag or touch it externally.

Obviously, whoever guesses the greatest number of knots wins.

4. Blind knots with a friend

This is a variation of the previous game. Divide the children into pairs. Give one of the pairs four cards with the names of the knots written on them and a piece of string long enough to tie them.

Blindfold one member of each pair while having the other choose a card and give him the rope. Once the knot has been tied, the blindfolded partner must try to identify it. The couple who guesses the most knots win.

You can also switch roles so that everyone is blindfolded.

5. Each knot is a step

Have all the children line up in the same row. In front of them place a kind of finish line. Give each of them a rope of at least 3 feet.

Pronounce the name of a knot. Each child will tie the knot. There must be a judge who comes by to check that the knot tied is correct.

Only if the knot is tied well can the child take a step forward.

The operation is repeated until one child - the winner - crosses the finish line first.

6. The human knot

This game can be played with up to 12 participants.

Everyone must stand in a circle. Each child must grasp the right hand of someone else who is not directly to their right or left. Then they must reach into the center of the circle and grasp someone else's hand with their free left hand.

Enjoy the spectacle, after all, hands have been grasped by others, of seeing a human knot!

Chapter 3

4 ORIGINAL FRIENDSHIP BRACELETS

1. Easy wave friendship bracelet

Materials

- 1 reel of embroidery cable in the color of your choice in the following length:
- 1 x 16-inch (40.64 cm) rope
- 1 x 24-inch rope (approx. 61 cm)
- 1 reel of embroidery cable in the color of your choice in the following length:
- 1 x 16-inch (40.64 cm) rope
- 1 x 24-inch rope (approx. 61 cm)
- macramé board or stand (optional)
- scissors.

Procedure

STEP 1: Take all 4 wires and fix them to a support with needles or tape.

STEP 2: Tie a knot with all 4 threads at about 3 inches (about 7.5 cm) below the tape.

STEP 3: Take the 2 pieces of 16-inch cable and stretch them down the center. Fasten the end to the surface.

STEP 4: Separate the other two cables of different colors and lengths, one to the right and one to the left of the central cables.

STEP 5: bend the left wire over the central wires and start a series of square knots (if you don't remember how to do this, look at book 2).

STEP 6: Repeat the knots until you reach the length you need (you can use your wrist as a tester).

Leave at least another 1.5-2 inches (about 5 cm) for the clasp and test the length again on your wrist.

STEP 7: tie a final knot and cut off the excess thread.

2. Chevron Friendship Bracelet

Materials

* 6 bobbins of embroidery thread of different colors with strings of the following lengths:
* 2 strings for each color 36-inch (91.44 cm) long
* macramé board or stand
* scotch
* scissors.

Procedure

STEP 1: Take all the threads and join them together with a wrapping knot.

STEP 2: Secure all knots to your work support with a piece of tape.

STEP 3: In total, you should have 12 working strings, 2 for each color.

Arrange the colors so that they mirror each other.

Take six different colored strings and place them all to the left.

Take the strings with the same 6 different colors and place them all to the right in the same position as those on the left.

STEP 4: Take the outermost wires on the left.

Hold the leftmost wire and place it in the shape of a 4 above the other wire. Wrap the wire underneath and pass the end through the ring (as in the photo).

Pull the end and tighten the knot.

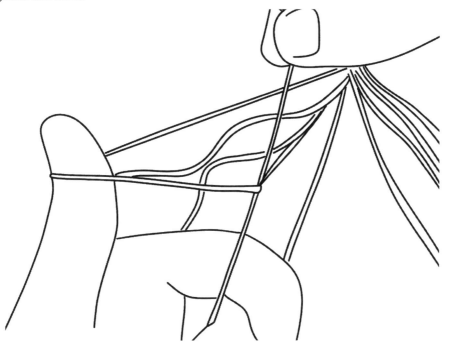

STEP 5: Repeat the operation one more time.

STEP 6: Still with the leftmost cable, tie the same knot with all other colors until you reach the center.

STEP 7: Repeat steps 4 to 6 with the cables on the right.

STEP 8: Take the two central cords and cross them from right to left. Wrap the end of the right cord around and across to make a backward knot and tighten it.

STEP 9: Repeat once more.

STEP 10: All the steps you have done so far complete one line.

Take the penultimate thread on the left and repeat the process on the right and left.

STEP 11: Repeat until you have finished all the colors.

STEP 12: Repeat the pattern until the bracelet is as long as you wish.

STEP 13: Close with a wrapping knot and cut off the excess threads.

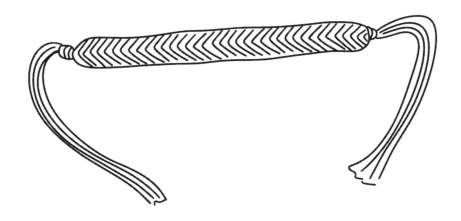

3. Easy colorful striped bracelet

Materials

- embroidery thread in various colors, 36-inch (91.44 cm) long. You can use 4 to 8 cables. In this case, 6 cables were used.
- adhesive tape
- scissors.

Procedure

STEP 1: Take all 6 threads and join them together with a wrapping knot.

STEP 2: Secure all knots to your work support with a piece of tape.

Stretch the strings in the order of the colors you desire.

STEP 3: Starting with the outermost string on the left, hold it in front of the next string forming a four.

Pass the rope through the loop and tie a knot.

STEP 4: knot the first rope in the same way with all the other colors.

STEP 5: Move on to the next cable and repeat the operation from left to right.

STEP 6: Repeat the same steps until the bracelet is the desired length.

Now gather all the threads into a knot and cut the extra string.

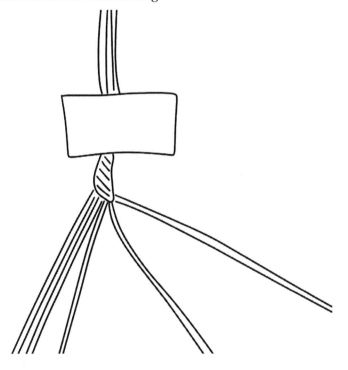

4. Celtic Friendship Bracelet

Materials

To make these bracelets, you can experiment with different materials, and use hemp or leather threads.

- 2 strands of hemp or leather. The length depends on how many Celtic knots you want to tie and the size of your wrist.
- scissors.

Procedure

To make a Celtic friendship bracelet, all you need to do is follow a simple pattern: make a series of Celtic knots, that are nothing more than Josephine knots in macramé (we saw this in book 2).

This bracelet is beautiful to give as a gift because the knot, recalling the infinity symbol, represents eternal love, devotion, and friendship.

STEP 1: Form a loop with the first thread and lay the other thread under the loop.

STEP 2: Take the thread underneath and slide it over the thread loop above.

STEP 3: Cross the thread that was underneath with the loop ends of the other thread.

STEP 4: Pass the wire that was underneath through the loop and tighten.

Follow the pattern for the Josephine Knot:

Chapter 4

3 TOP KIDS-FRIENDLY PROJECTS

1. Rainbow macramé

The rainbow macramé is one of the most adorable projects you can do with macramé.

You can make one, following the step-by-step instructions here with your children, and use it to decorate their bedroom.

They are truly fashionable!

Materials

* 25 mm (approx. 1 inch) single-strand cotton (or other available material) macramé rope, cut to the following lengths:
* 1 x 7-inch rope
* 1 x 9-inch rope
* 1 x 11-inch rope
* 1 x 14-inch rope
* 4 yarns, each of a different color (we are making a rainbow!)
* 1 piece of thick felt, approx. 6.5-inch x 8.5 inch (16.51 x 21.6 cm)
* 1 piece of 7-inch (approx. 18 cm) long string for the hook (you can use whatever material you have on hand)
* scotch
* hot-melt glue gun with a glue stick
* scissors
* fringe brush.

Procedure

STEP 1: Secure the ends of the thicker ropes you cut with tape to prevent them from fraying.

STEP 2: Take one of the thicker cords and cover it with the different colored yarn.

Hold one end of the cable and simply wrap it with the colored wire until it has covered its entire length.

STEP 3: Repeat the procedure for the other 3 strings.

STEP 4: Once all the thick ropes have been wrapped, it is time to glue them onto the felt cloth.

First, do a test. Place all the strings on the felt (from the smallest to the largest) and check their correct position.

Start with the smallest band. Add hot glue to the back and press it onto the felt so that the glue cools.

STEP 5: Add the piece of string that serves as a hook in the center.

STEP 6: Proceed to glue the other pieces of rope.

STEP 7: Remove the tape from the ends of the cables and even them out.

STEP 8: Turn the rainbow upside down and carefully cut off the excess felt.

STEP 9: Brush the ends of the strings to form the fringe.

Here is your beautiful macramé rainbow!

2. Macramé dream catchers

Dreamcatchers can come in many shapes and sizes. Here we propose a quick and easy one to make with your kids, in the shape of the moon, which you can hang in their rooms to chase away bad dreams.

Materials

- 12-inch (30.48 cm) long crescent frame
- 3 mm braided cotton cord in black (or another color) in the following sizes:
- 50 x 63-inch (160 cm) ropes
- 9 x 45-inch (114 cm) strings
- 40 x 8-inch (20.32 cm) strings
- scissors.

Procedure

STEP 1: Take the half-moon stand and hang it on a support so that the inside faces the other way.

STEP 2: Attach the first 9 45-inch strings with an inverted lark's head knot. Move them to the left and attach 7 more 63-inch strings on the right side.

STEP 3: Take the first rope to the left and do a DHHK around the moon-shaped frame.

STEP 4: Make a square knot using the rope you have just attached to the frame and the next 3 ropes.

STEP 5: Repeat with the right ropes.

STEP 6: Fasten 2 left-hand ropes to the frame with other DHHK.

STEP 7: Skipping the 2 ropes you attached to the frame, tie a row of square knots alternating with the first one.

STEP 8: Make a subsequent row of square knots alternating with the previous one by skipping two more ropes to the left and right.

STEP 9: Repeat until you end up with a row of only 2 square knots (see photo).

STEP 10: Pull all cables behind the frame.

Attach each string to the frame with the DHHK.

Tighten well so that the square knot pattern is clearly visible.

Attach the first 2 strings on the left then move to the right, to check that the pattern tension is even.

STEP 11: Cut the left-hand cables while leaving the right-hand cables long.

STEP 12: Rotate the dream catcher so that it 'stands' with the longer wires hanging down.

STEP 13: Tie 14 more 63-inch ropes at the bottom with lark's head knots. Two ropes between each knot.

STEP 14: Take the 4 center strings and tie a square knot.

Make 3 square knots left of center and 3 square knots right of center.

Make another 6 rows of alternating square knots, always skipping 2 strings at the beginning and end of the row (see photo below).

STEP 15: Take the third and fourth cables from the outside on both sides. Pull them to the center and tie a square knot just below the last square knot you tied just now.

STEP 16: Take 28 63-inch strings and attach them with a DHHK to the string, 14 on the left and 14 on the right.

STEP 17: Join the center strings with a final DHHK and you're done!

3. Fish-shaped key ring

We chose this type of pattern because it is the one most used to make most key rings.

Materials

- 3mm (0.12 inch) single-strand cotton macramé rope in the following lengths:

- 3 x 48-inch (122 cm) strings
- 1 x 20-inch (50.8 cm) rope
- key fob
- fringe brush
- Macrame Board or other support
- T-pins
- scissors.

Procedure

STEP 1: Take the keyring clasp and attach the three 48-inch strings with a lark's head knot.

STEP 2: Make 2 diagonal knots with DHHK starting from the left towards the bottom right.

STEP 3: Start from the right and tie 3 DHHK knots from right to left at the bottom.

STEP 4: You will see a V-shaped pattern forming.

Continue for 6 more lines.

STEP 5: You have reached the eighth row.

From the left, do a DHHK diagonally downwards to the right.

Hold the string tight and do the same thing on the right.

STEP 6: Hold all the strings together.

Take the last 20-inch rope and finish with a wrapping knot.

Cut off the ends of the extra rope.

STEP 7: Comb out the length of the remaining strings with the comb.

Cut the fringe thus created to give it the shape of a fishtail (basically an inverted V).

CONCLUSIONS

3 final tips

We have come to the end of this wonderful journey to discover all the secrets of macramé. We have explored so many aspects of this original art form that you have surely already tasted how much emotion it can convey.

And still, there is so much to learn and make with your own hands. Even if you don't become a macramé superstar, follow these last 3 tips:

1. Keep practicing while having fun.
2. Share your passion with others and, why not, teach them what you have learned.
3. Don't just follow the patterns or tutorials of others to the letter, develop your creativity and draw inspiration from what you like.

Scan this QR code or type in the link below to get access to your

Easy Download and Print Instructions with Extra Pictures for all the Projects in this Book

No signup required!!

https://drive.google.com/drive/u/0/folders/15rUBULA7NmjxCTZ46fxAxwSBK_f2851Y

Made in the USA
Coppell, TX
03 February 2024

28505854R00092